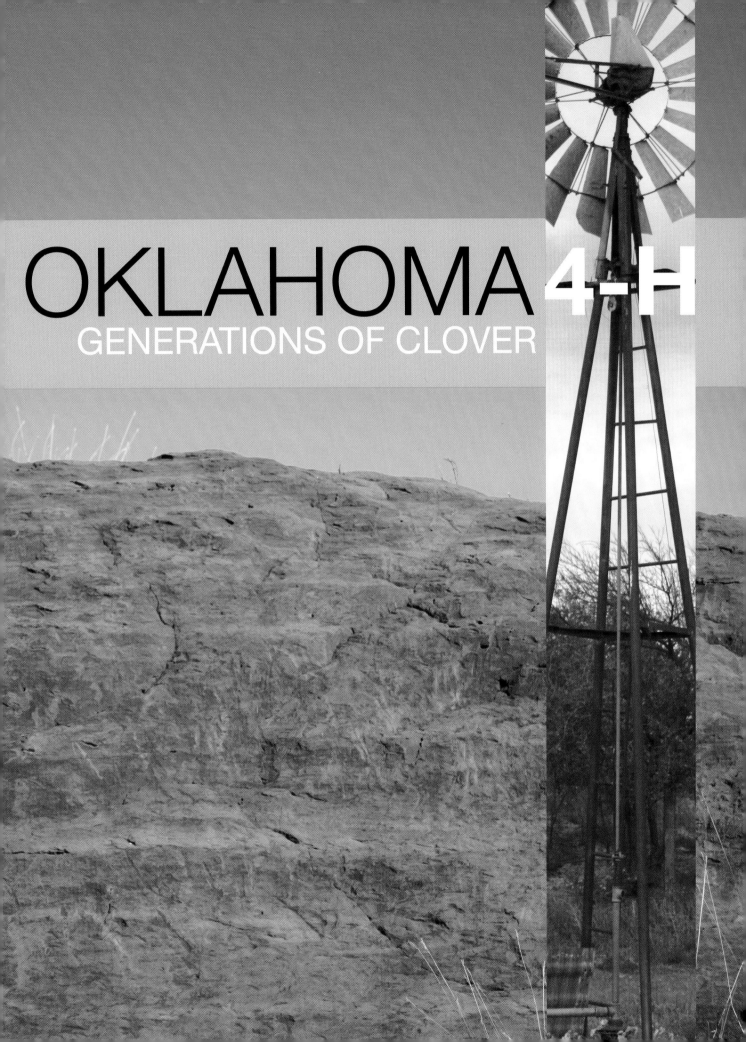

OKLAHOMA 4-H
GENERATIONS OF CLOVER

BY JESSICA STEWART
AND CAITLIN SCHEIHING

Copyright © 2010 by the Oklahoma State 4-H

All rights reserved, including the right to reproduce this work in any form whatsoever without permission in writing from the publisher, except for brief passages in connection with a review. For information, please write:

The Donning Company Publishers
184 Business Park Drive, Suite 206
Virginia Beach, VA 23462

Steve Mull, General Manager
Barbara Buchanan, Office Manager
Heather L. Floyd, Editor
Chad Casey, Graphic Designer
Derek Eley, Imaging Artist
Cindy Smith, Project Research Coordinator
Tonya Hannink, Marketing Specialist
Pamela Engelhard, Marketing Advisor

Wade Grout, Project Director

Cataloging-in-Publication Data
Library of Congress Cataloging-in-Publication Data

Stewart, Jessica, 1983-
  Oklahoma 4-H : generations of clover / by Jessica Stewart and Caitlin Scheihing.
       p. cm.
  Includes bibliographical references.
  ISBN 978-1-57864-614-2 (hardcover : alk. paper)
 1. 4-H clubs--Oklahoma--History. 2. Youth--Oklahoma--Societies and clubs--History. 3. Education--Oklahoma--History. 4. Youth development--Oklahoma--History. 5. Agricultural education--Oklahoma--History. I. Scheihing, Caitlin, 1986- II. Title.
  S533.F66S85 2010
  369.4--dc22

                               2009052236

Printed in the United States of America at Walsworth Publishing Company

Table of Contents

7	Preface
8	Foreword
10	Celebrating the Oklahoma 4-H Centennial: A Snapshot
12	Celebrating the Oklahoma 4-H Centennial: Centennial Families
14	Gallagher-Iba: The Madison Square Garden of the Plains
16	History of the Oklahoma Association of Extension 4-H Agents
17	Langston University 4-H Youth Development
18	Oklahoma 4-H: 50th Anniversary Marker
19	Oklahoma 4-H: Ambassadors
20	Oklahoma 4-H: Citizenship Washington Focus
21	Oklahoma 4-H: Collegiate 4-H
22	Oklahoma 4-H Foundation… A Growing Legacy
23	Oklahoma 4-H: Key Club
24	Oklahoma 4-H: National Congress
27	Oklahoma 4-H: County Histories

28	Adair County	51	Cimarron County	68	Grant County
30	Alfalfa County	51	Cleveland County	70	Greer County
31	Atoka County	52	Coal County	71	Harmon County
34	Beaver County	54	Craig County	72	Harper County
37	Beckham County	57	Creek County	72	Haskell County
39	Blaine County	57	Custer County	75	Hughes County
41	Bryan County	58	Delaware County	77	Jackson County
43	Caddo County	60	Dewey County	78	Jefferson County
45	Canadian County	62	Ellis County	78	Johnston County
47	Carter County	64	Garfield County	81	Kay County
48	Cherokee County	66	Garvin County	84	Kingfisher County
50	Choctaw County	67	Grady County	87	Kiowa County

(Table of Contents continued on next page)

(Table of Contents continued from previous page)

90	Latimer County	115	Noble County	146	Roger Mills County
92	LeFlore County	118	Nowata County	148	Rogers County
92	Lincoln County	120	Okfuskee County	149	Seminole County
95	Logan County	121	Oklahoma County	151	Sequoyah County
97	Love County	124	Okmulgee County	153	Stephens County
98	Major County	126	Osage County	155	Texas County
100	Marshall County	128	Ottawa County	157	Tillman County
104	Mayes County	132	Pawnee County	159	Tulsa County
106	McClain County	136	Payne County	161	Wagoner County
108	McCurtain County	140	Pittsburg County	163	Washington County
110	McIntosh County	140	Pontotoc County	164	Washita County
111	Murray County	143	Pottawatomie County	166	Woods County
114	Muskogee County	145	Pushmataha County	167	Woodward County

170	Oklahoma 4-H Hall of Fame
172	Oklahoma 4-H State Presidents
174	Oklahomans in the National 4-H Hall of Fame
174	Oklahoma 4-H State Leaders
175	Bibliography
176	About the Authors

Preface

Contributed by Dr. Eugene "Pete" Williams

Celebrating one hundred years of 4-H work provides the opportunity to review and analyze the program in Oklahoma.

The 4-H organization, as a part of the Oklahoma State University Cooperative Extension working in partnership with the Oklahoma 4-H Foundation, the U.S. Department of Agriculture, and the National 4-H Council, has focused on the positive development of youth through educational programs. Oklahoma 4-H members and leaders have been busy learning and growing through projects and activities for these past one hundred years.

Observing the Oklahoma 4-H Program as a 4-H member and as an Extension worker at the county, state, and national level, I'm convinced that Oklahoma has a program that is second to none. The Oklahoma 4-H delegates, each with a trademark white scarf at the National 4-H Congress, exhibit pride in their many achievements and national awards. Delegates from other states are inspired by Oklahoma's traditions.

It is with a great deal of pride that I review the progress and development of 4-H work in Oklahoma in its first one hundred years. Not only has 4-H been a great program for the development of the youth of the state; it has had an equal impact on the development of adult volunteers in every neighborhood within the state. The 4-H members, their parents, and the adult volunteer leaders continue to provide at the community, county, state, and national level.

We are pleased to celebrate one hundred years of 4-H work in Oklahoma with the 4-H Centennial Homecoming in November 2009 and the publication of this 4-H history book. The comments and statements of former 4-H members in this book of memories are evidence of the impact that the Oklahoma 4-H Program has made in their lives.

Thank you, Centennial Committee, for the organization of the 4-H Centennial celebration and the development of this 4-H history book. We hope this book's collection of memories from the past serve as an inspiration for your continued interest in and support of Oklahoma 4-H.

Foreword

Contributed by Dr. Charles Cox

As Oklahoma 4-H celebrates its one hundredth birthday, we are pleased to present this book as a remembrance of a special time of honoring and celebrating our history, current programs, and the people of Oklahoma 4-H while beginning to envision what the future might hold. As the Oklahoma 4-H Program leader, I work in support of field staff, 4-H members, and volunteers across our state and get to see and hear thousands of stories about 4-H. The many testimonials indicate that 4-H has positively impacted the lives of millions of young people in Oklahoma and around the world. It is safe to assume that W. D. Bentley, the founder of the Extension program in Oklahoma, would be amazed at what has grown out of the corn and tomato clubs of 1909.

All these years later, 4-H members and volunteers are busy learning and growing as they work and play. Approximately 125,000 youth are involved in educational programs and positive youth development led by Extension educators working with more than 6,000 volunteers. Within the pages of this book, one will find information and photos from 4-H's past and present. Each unique story and face represents a small piece of the 4-H story in Oklahoma. In small towns, on rural farms, and in metropolitan communities across the state, memories have been made and lives have been changed as a result of caring adults sharing their lives with 4-H youth. We are blessed to live in a state whose citizens value and consistently invest in the future by supporting 4-H.

Obviously, this book only captures a small representation of the people and events that shape the history of 4-H in Oklahoma. Realize that a large number of heroes and heroines are not in this book, yet they live vividly in the hearts, minds, and memories of former and current 4-H members. We are as old as one hundred years and as current as today. The 4-H program continues to evolve to meet the educational and emotional needs of our youth. Oklahoma 4-H, as part of the Oklahoma State University Cooperative Extension Service, working in partnership with Langston University, the Oklahoma 4-H Foundation, the U.S. Department of Agriculture, and the National 4-H Council, appreciates your support. Please continue to find ways to be involved in supporting youth and helping them build their own memories. If you have not been involved in 4-H since you were a youth, consider what you have to give back. Perhaps you should consider becoming a volunteer or perhaps you need to involve your own children or grandchildren in 4-H.

We thank the educators, volunteers, and specialists who have worked to make this book possible with their stories and photographs. We hope this book brings a smile to your face and serves as an inspiration for your continued interest in 4-H. It has been, as our Centennial Committee eloquently states, a wonderful time to celebrate one hundred years of Oklahoma 4-H: Honoring. Celebrating. Envisioning.

9

CELEBRATING THE OKLAHOMA 4-H CENTENNIAL

CENTENNIAL
1909-2009

A Snapshot

Project: 4-H Centennial Garden
Planning time: July 2008–July 2009
Members: Cloverbuds and 4-H'ers from Payne, Noble, Pawnee, Tulsa, and Adair counties
Mission: To create a 4-H Centennial Garden at the Oklahoma Botanical Gardens

More than twenty youth and their parents, along with Adair County Extension Educator Nancy Arnett and Oklahoma gardening hostess Kim Rebek, designed, developed, and planted the Oklahoma 4-H Centennial Garden. A commemorative garden celebrating 4-H from its corn club days to the present, the garden featured corn, tomatoes, okra, Oklahoma-native flowers, and a solar-powered fountain. Team members learned how teamwork and leadership create results, no matter how large the project seems.

Project: 4-H Centennial Road Trip
Planning time: January 2008–December 2008
Members: 4-H members, 4-H volunteers, 4-H families, and Extension educators
Mission: For 4-H members and volunteers to explore the great state they live in while becoming knowledgeable of their 4-H heritage during the 4-H Centennial celebration

Oklahoma 4-H values the importance of celebrating the people, places, and events that have shaped our organization and state. Youth and adults piled into cars, vans, and even on horseback in an effort to honor, celebrate, and envision the practice of leadership, citizenship, and life skills development on the 4-H Centennial Road Trip. Members increased their 4-H heritage knowledge and became geographically acquainted with the state as they worked in partnership with adults through the process of planning and organizing their trip.

Project: Eskimo Joe's 4-H T-Shirt
Planning time: October 2008–April 2009
Members: Oklahoma 4-H and the Eskimo Joe's Team
Mission: To create a commemorative Eskimo Joe's 4-H Centennial T-Shirt

When Oklahoma 4-H staff approached Eskimo Joe's with the idea of having an Eskimo Joe's 4-H Centennial T-Shirt, they didn't know it would become such a success and result in the first-ever Eskimo Joe's T-Shirt Release Party! With more than 300 shirts sold on release day, March 6, 2009, an estimated 1,500 shirts sold within the first three months of the shirt's release. Thanks to the generosity of Eskimo Joe's, a portion of the proceeds has been donated to Oklahoma 4-H to endow a scholarship for Oklahoma 4-H youth.

Project: 4-H Corn Maze
Planning time: Summer 2008–Fall 2009
Members: Oklahoma 4-H and P Bar Farms
Mission: To create a 4-H Centennial corn maze

Loren Liebscher of P Bar Farms created an amazing seven-acre corn maze for Oklahoma 4-H to commemorate its centennial year. To add authenticity to the maze, Loren even planted it in corn to honor the corn clubs from which 4-H began. In October 2009, Oklahoma 4-H hosted a centennial party to celebrate one hundred years of Oklahoma 4-H. Oklahoma 4-H'ers from all over the state celebrated in 4-H style with games, crafts, and a chuck wagon dinner.

Project: 4-H Green Tie Gala
Planning time: Fall 2007–Fall 2009
Members: Oklahoma 4-H
Mission: To host a homecoming celebration

The finale event of the Oklahoma 4-H Centennial, the Green Tie Gala was hosted at the Skirvin-Hilton Hotel, chosen for its connection to 4-H in the 1940s and 1950s when 4-H'ers stayed at the Skirvin before getting on the Santa Fe train to head to Chicago Club Congress. The Green Tie Gala was an event all its own, for Oklahoma 4-H had never hosted such an affair. Guests enjoyed a showcase of 4-H talent and programs envisioning the next one hundred years of Oklahoma 4-H.

Project: 4-H Centennial Fair Classes
Planning time: 2007–2008
Members: Oklahoma 4-H district program specialists and state 4-H staff
Mission: To create ways for 4-H members and alumni to showcase their 4-H heritage

Special to the centennial year, county and state fairs offered 4-H members and alumni centennial classes in 2008, 2009, and 2010. With more than 200 exhibits selected for the 2008 Oklahoma City and Tulsa state fairs, Oklahoma 4-H'ers and alumni showed how 4-H has positively affected their lives. From original project work to stories about 4-H across generations, Oklahoma 4-H history came alive.

Centennial Families

2009 Adult Recipients

Ruth Ann Givens

A fourth-generation 4-H alumna, Ruth Ann Givens says she has strongly encouraged 4-H involvement because of her memories of her 4-H experiences.

"[The] 4-H [program] taught me to strive for your best in whatever your endeavor through the positive competition it provides," Ruth Ann said. "I learned that being a leader is your responsibility if you have the knowledge and capability to do so."

Ruth Ann says her family bleeds green, and as a result, her family will continue to promote the 4-H program.

"I truly believe this drive and desire to push oneself for excellence has been instigated and nurtured by the 4-H program in our family," Ruth Ann said.

John H. Pfeiffer Jr.

A fourth-generation 4-H alumnus, John H. Pfeiffer Jr. says the skills he learned in 4-H gave him the confidence to build a life that centers on community service and working with young people.

"Without a doubt, the 4-H experience can be summed up in one word: involvement," John said. "When you are nine, you just want to be a part of the group that is going to camp, or giving a speech, or showing a lamb. In a very short time, you realize the more involved and engaged you are in the process, the more fun you are having and the more success you attain."

John says his family sees 4-H as the most significant way to get youth started on the right track.

"We feel it's important for them to join because it starts the process of building the community leaders who are critical for our small communities to thrive."

Annette Stowers

With five generations of 4-H in her family, Annette Stowers says her life has come full circle from being an active 4-H member in Woodward County to now being a certified volunteer and club leader of the Trailblazers Horse Club in Cleveland County.

Annette says 4-H will continue to influence future generations because there's no limit to what 4-H can do. She added, "4-H has developed from just teaching the basic life skills of canning and planting crops to teaching science and technology skills by having robotic clubs and teaching computer skills."

Although the basic goal of 4-H hasn't changed, Annette can only imagine what her great-grandmother would think of 4-H now.

"Although the basic life skills are still being taught, we are now making robots, traveling all over the United States for leadership conferences, and using technology in so many new ways. How times have changed!"

Dea Rash

With four generations of 4-H in her family, Dea Rash says she isn't sure exactly how 4-H became a tradition because it's always been a part of her family.

Dea said, "4-H has helped four generations of our family members learn valuable life skills such as public speaking, citizenship, leadership, recordkeeping, teamwork, and much more."

As a 4-H parent and a Family Consumer Sciences/4-H Extension educator, Dea says she truly appreciates the sacrifices that her family made for her to make sure she got the most out of her 4-H experience.

"Through our continued involvement throughout many more generations, the rich 4-H heritage that we have built will continue to grow," Dea said.

2009 Youth Recipients

Austin Rhye Kindschi

A third-generation 4-H'er, Austin Rhye Kindschi says 4-H has taught him that he can achieve anything he sets his mind to.

"When I turned nine, there was no question that I would join 4-H," Austin said. "My grandfather, my mom and dad, nearly all of my uncles, and so many cousins I can't count them all, were in 4-H."

Austin says Oklahoma 4-H will continue to influence future generations because it has adapted to the changing needs of Oklahoma's youth.

"[The] 4-H leaders are willing and able to move with the times and adapt in ways that still keep kids interested in learning and growing the 4-H way," Austin said. "Through 4-H, I have learned many skills that will help me choose a career path that will enable me to give back to the organization that has given and taught me so much."

Cheyne Sierra Wright

A third-generation 4-H'er, Cheyne Sierra Wright says 4-H has affected her family in many positive ways. Her parents are both 4-H Shotgun Club instructors and her grandparents help with 4-H projects, too.

"[The] 4-H [program] has shown me where I have skills, has taught me a lot of them, and has given me an idea of where I want to be when I grow up," Cheyne said. "Since I can see how much 4-H has helped me and my brothers, I know it has been an influence on our generation and I will teach it to my children someday, too."

Cheyne says Oklahoma 4-H continues to influence future generations because "4-H adds new projects to keep up with technology that allows kids with different interests to belong." Cheyne added, "4-H teaches kids to follow through on projects, responsibility, and hard work. The best thing about 4-H is it is something that our whole family does together. It's been that way for three generations of our family and it just keeps getting better!"

Aaron Sharp

A fourth-generation 4-H'er, Aaron Sharp's family has been involved in 4-H since the 1920s. Aaron says 4-H started with his great-grandpa, who was a 4-H leader, and now each child in the family joins 4-H as soon as they are of age.

"Seeing the friendships that develop and grow within the 4-H program has made the biggest impact on me," Aaron said. "It's a small, small world and this proves to me that the 4-H program is a powerful life tool."

From Aaron's great-grandpa's leadership abilities to his uncle's national poultry recognition, 4-H continues to be a strong influence on the Sharp family. Aaron's parents continue to be involved in 4-H, and his older sister, Alea, is a 4-H alumna.

"It seems that the work invested as a 4-H member grows and develops throughout the years," Aaron said. "What a legacy!"

Taler Sawatzky

A third-generation 4-H'er, Taler Sawatzky says she wouldn't be where she is today without 4-H.

"[The] 4-H [program] has opened the doors to so many leadership and citizenship projects and programs, and has really made me aware of what responsible citizenship means," Taler said. "I plan to continue being involved in 4-H after I graduate by being a member of the alumni association, as well as being a volunteer leader."

In the Sawatzky family, 4-H began with Taler's grandmother, who was a 4-H member in Caddo County. Taler says with her mom being an Extension educator, she didn't have a chance not to be in 4-H.

"It has been an honor to call myself a 4-H member and to be elected by my peers to serve as the state 4-H secretary. [The] 4-H [program] is a tradition in our family and someday my kids will be in 4-H."

Gallagher-Iba: The Madison Square Garden of the Plains

Adapted from a story by Jan Garms

Gallagher Hall, Gallagher-Iba, and even the "Madison Square Gardens of the Plains" are what people have called the famous arena on the Oklahoma State University campus, but many people do not realize this building was dedicated and once called the 4-H Club and Student Activities Building.

In the early years of 4-H Roundup, students would either stay in tents or in the homes of faculty and citizens of Stillwater. However, in the late 1930s, the tents used for the meetings collapsed in a thunderstorm and injured some of the youth.

Oklahoma A&M College President Henry Bennett used this accident to go to the legislature and request funds for a 4-H meeting hall. Bennett knew he needed a new building for his basketball and wrestling programs. He also knew the 4-H group was the largest statewide group using the campus at the time, and they represented young people from all over the state.

Within the House Bill, written in 1938, Bennett seized the opportunity to name the building the 4-H Club and Student Activities Building. When Bennett went to the legislature, he argued that the school needed money to build a building to hold events such as the State 4-H Roundup on the OAMC campus.

According to Oklahoma State University's Centennial Histories series, "An unusual provision on the main floor was the original installation of two huge curtains which could be rolled up or down against ceiling hangers. Their intended use was to separate the central core of the building into three parts so separate district 4-H club meetings could be conducted at the same time." Although used for other activities in addition to 4-H, Bennett was always careful to refer to the building as the 4-H Club and Student Activities Building, and on June 1, 1939, the building was officially dedicated during the State 4-H Roundup as the 4-H Club and Student Activities Building. However, later in 1939, the building was christened and dedicated to the wrestling coach, Edward Gallagher. From that time on, students, alumni, and the media referred to the building as Gallagher Hall.

Subtle touches of Gallagher-Iba's history are still shining through even with the renovations that have been done over the years. The original stone mural that decorated the outside of the 4-H building can still be seen on the arena walls as well as the 4-H case displaying a green jacket that sheds light on the past of the historic building.

Gallagher-Iba: The Madison Square Garden of the Plains

History of the Oklahoma Association of Extension 4-H Agents

Contributed by Susan Murray

The idea for a state professional association of 4-H professionals began in the early 1970s and blossomed in 1974, when several state and county staff members attended the National Association of Extension 4-H Agents Conference in Wichita, Kansas. Seeing the need for a state association, these staff members returned to Oklahoma and laid the groundwork for the Oklahoma Association of Extension 4-H Agents. Under the guidance and leadership of Sheila Forbes, OAE4-HA was formally organized in May of 1975 at the State 4-H Roundup. NAE4-HA President Nancy Ascue attended the organizational meeting and gave leadership to the association. The new association had the full support of state administrators including Associate Extension Director William F. Taggart and State 4-H Program Leader Eugene "Pete" Williams, who were charter members.

A total of twelve members attended the 1975 NAE4-HA Conference in Louisville, Kentucky. Their ideas and creativity helped bring about the first OAE4-HA Conference, held on the Oklahoma State University campus in July of 1976. The theme for the first conference was appropriate for the occasion and for the year: "Proclaiming Professionalism."

At the NAE4-HA Conference in Denver, OAE4-HA successfully bid against Texas to host the 2004 NAE4-HA Conference in Oklahoma City. More than one hundred OAE4-HA members, life members, and volunteers successfully hosted this meeting for more than 1,200 members and guests.

Several OAE4-HA members have held elected or appointed positions on the NAE4-HA Board:

Barbara Hatfield	Southern Region Director, 1991–1993
	NAE4-HA Member Recognition Chair, 1997–1999
	2004 NAE4-HA Conference Co-chair 2000–2004
Mike Klumpp	Southern Region Director, 1987–1989
Tomas Manske	Southern Region Director, 2007–2009
Susan Murray	2004 NAE4-HA Conference Co-chair 2000–2004
David Sorrell	Southern Region Director, 1997–1999
	Vice President for Marketing and Outreach, 2004–2006
Pat Trotter	Southern Region Director, 1979–198
	NAE4-HA Vice President, 1981–1982
	NAE4-HA President-elect, 1982–1983
	NAE4-HA President, 1983–1984

In addition, Sheila Forbes, Merl Miller, Mickey Prescott Simpson, Charles Cox, Jim Rutledge, Susan Meitl, and Kyle Worthington have served in regional leadership positions with NAE4-HA committees.

Three OAE4-HA members are recipients of the prestigious American Spirit Award:
1985 – Eugene "Pete" Williams
1996 – Mary Sue Sanders
2006 – Jim Rutledge (as an Oregon life member)

The following OAE4-HA life members were inducted into the National 4-H Hall of Fame:
2002 – Mary Sue Sanders
2003 – Eugene "Pete" Williams
2006 – Ray Parker
2009 – Joe Hughes
2009 – Barbara Hatfield

Langston University 4-H Youth Development

Contributed by Dorothy Wilson

From its start, the Oklahoma 4-H clubs have embraced youth of all racial groups. Throughout the early 1900s the 4-H clubs for African American youth focused not only on improving agricultural practices but also on wildlife, canning, and even automobile driving. By 1912, Oklahoma had also employed eleven women agents, including African American Annie Peters Hunter of Boley.

While there were special agents to work with African American youth and later to work with Indians, it was not until the Civil Rights Act of 1964 that the integration of 4-H was ensured. Throughout the stories of this book, we see records of historical work in both the segregated and the integrated 4-H clubs.

The Langston University Cooperative Extension 4-H Youth Development Program was officially initiated in 1988 and continues to recruit minority youth and other youth into organized 4-H clubs. In 2005, the Agricultural Research, Education, and Extension Complex opened, which houses the 4-H Youth Development Program. New programs were organized with an emphasis on establishing 4-H goat and fish clubs in underserved areas. At the same time, new school enrichment programs, including the 4-H Goat Kid in the Classroom and the 4-H Aquatic School Enrichment programs, were established in several Oklahoma counties. These 4-H programs were instrumental in educating youth and adults in underserved communities. As the Langston University Cooperative Extension 4-H Youth Development Program has grown, 4-H youth have participated in 4-H Day at the Capitol since 2001. In 2008, Langston University 4-H member Joshua Stevenson was selected as a page for the Honorable Representative Lee Denney and the first spokesperson for Langston University during 4-H Day at the Capitol.

With an emphasis by Langston staff, 4-H members recruited by the Langston University 4-H Program participated and won awards at the county dress revue contest, county speech contest, county and state fairs, and other 4-H events conducted at the local, state, and national levels.

Since 1988, the Langston University Cooperative Extension Service has sent delegates to participate in several national events. Langston University Cooperative Extension selected its first delegates to attend the 59th Annual National 4-H Conference in Washington, D.C., and the 68th National 4-H Congress in Chicago, Illinois.

In 2004, Dr. Ernest L. Holloway, Langston University president, was inducted into the National 4-H Hall of Fame at the National 4-H Conference Center in Chevy Chase, Maryland.

As we enter the second century of 4-H in Oklahoma, the membership mirrors the state in many ways in diversity. More than 26 percent of the youth involved in 4-H are minorities. The 4-H club program continues to be an integral part of both Langston University and Oklahoma State University.

The leaders established a great legacy of education that continues in the joint efforts of land-grant universities as they are busy promoting positive youth development opportunities available to all youth.

Oklahoma 4-H: 50th Anniversary Marker

Contributed by Vickie Luster

In 1959, Oklahoma 4-H celebrated its fiftieth anniversary with a crowd of nearly 2,000 people from across the state coming to Tishomingo for a celebration.

The celebration began at 10 a.m. with commemorative ceremonies in Beames Field House on the campus of Murray State College. Three of Oklahoma's top political figures and hundreds of youth and their leaders were on hand for the program and the dedication of the granite marker. Vera Taylor, longtime Johnston County Extension home economist, stated in an interview that 4-H clubs from around the state brought a small bag of dirt to the location of the marker to show unity in Oklahoma 4-H.

Guest speakers were United States Congressman Carl Albert, United States Senator Robert S. Kerr, and Oklahoma Governor Ed Edmondson. Oklahoma State Extension Director Luther Brannon was also present for the day's activity.

Despite the December 16th wintery weather, the crowd gathered at the Chickasaw National Capital Building, which was serving as the Johnston County Courthouse for a special dedication of a granite marker. Participating in the program that day were former 4-H'ers Stanley Anderson, Donnie Rowland, and Lewis Parkhill. Anderson became an attorney and later a county judge; Rowland is a retired schoolteacher and administrator; and Parkhill is a retired college professor and currently serves as mayor of Tishomingo.

The granite monument has a bronze plaque with a 4-H Clover surrounded by the words "HEAD," "HEART," "HANDS," and "HEALTH," and reads as follows: "A half-century of 4-H work. Oklahoma's first corn club of fifty boys, forerunner of today's 4-H program, was originated at Tishomingo in 1909 under the direction of W. D. Bentley, father of Extension Work in Oklahoma. In 1910, tomato canning clubs were formed for girls. Fifty years later, 61,650 boys and girls were enrolled, making 4-H one of the largest youth organizations in Oklahoma. [The] 4-H members carry out supervised projects and participate in activities that develop leadership, citizenship, and a satisfying home life. County Extension agents, working through volunteer local leaders, guide these boys and girls to fulfill their motto: To Make the Best Better."

The monument was moved to the new Johnston County Courthouse complex at 403 West Main Street, Tishomingo, where it sits as a lasting tribute to the "grass roots" beginning of Oklahoma 4-H.

Oklahoma 4-H: Ambassadors

Contributed by Cathy Allen, Shannon Ferrell, Mary Sue Sanders, Gwen Shaw, and Jessica Stewart

Envisioning. The word, used throughout the Oklahoma 4-H Centennial, has prominence in the history of Oklahoma 4-H. W. D. Bentley envisioned the first 4-H club in 1909. Oklahoma A&M President Henry Bennett envisioned a 4-H Club and Student Activities Building in 1938, and Mary Sue Sanders envisioned the Oklahoma 4-H Ambassador Program in 1993.

Modeled after the National 4-H Ambassador Program, planning for the state program began in 1993 with the leadership of Mary Sue Sanders, Shannon Ferrell, Jim Rutledge, and Vernon McKown. By 1994, the State 4-H Ambassador Program was born. Of twenty-three applicants, fourteen were accepted.

Although a relatively young program, the 4-H Ambassador Program already runs deep with tradition. State 4-H Ambassador candidates apply for consideration in early March, interview in May, and are announced at the annual State 4-H Roundup. Beginning in 1994, the 4-H Ambassador team began attending the 4-H Ambassador Retreat at the Bockelman Ranch. They, along with the Lavertys and Sherrers, host a weekend-long retreat for the team focusing on developing leadership characteristics, networking with donors, and creating their 4-H story.

To differentiate the State 4-H Ambassador team from the State 4-H Leadership Council (previously known as the State Officer team), the first 4-H Ambassador team chose to wear green jackets with navy skirts and pants. In addition, the first team asked Derrick Ott to design a crest to wear on the green jacket. Ambassadors proudly display their crests to represent the group at 4-H events.

In addition, in 1995, a special plaque was designed for ambassadors who had completed a year of the program. First-year ambassadors receive the plaque, while those who have served for a longer period are given discs to add to the plaques they have already received. In 1996, the plaques were given in memory of Dana Smith, a 4-H ambassador who was killed in a car accident.

Although the executive director of the Oklahoma 4-H Foundation has historically been the state contact for the 4-H Ambassador Program, in 2007, Cathy Allen, state 4-H curriculum coordinator, was appointed to the role. Volunteer advisers Gwen Shaw and Shannon Ferrell continue to assist in directing the team and planning activities with the team to improve leadership, citizenship, and donor relationship skills.

Whether ambassadors are sharing their 4-H stories with others, fundraising, speaking on the House and Senate floor on 4-H Day at the Capitol, or collaborating on new service project ideas, state 4-H ambassadors continue the tradition of serving as the state's top 4-H representatives. The success of the program relies on its advisers and the more than 193 4-H'ers who have served as ambassadors. They have shown that the 4-H Ambassador Program continues to be a strong tradition in the history of Oklahoma 4-H.

Oklahoma 4-H: Citizenship Washington Focus

Contributed by Charles Cox and Tracy Branch-Beck

Citizenship Washington Focus (CWF) is a 4-H leadership program for high school youth. For seven weeks out of the summer, delegations of fifteen- to nineteen-year-olds from across the country attend this six-day program at the National 4-H Youth Conference Center located just outside Washington, D.C., in Chevy Chase, Maryland. At one time, the program was known as the Citizenship Short Course, and before that, an event was held in D.C. called National 4-H Camp.

Oklahoma 4-H members are given the opportunity to explore, develop, and refine the civic engagement skills they need to be outstanding leaders in their home communities and at the national level.

Through sightseeing tours in the living classroom of Washington, D.C., and hands-on educational workshops, youth learn about the history of our nation, the leaders who have shaped it, and how they can apply the leadership and citizenship skills they have learned at CWF when they return home.

Oklahoma 4-H youth and chaperones travel my chartered bus making stops along the way at historical points of interest. Over the years, many lifelong friendships have been made on this trip—and even a few romances!

Oklahoma 4-H: Collegiate 4-H

Contributed by Melanie Skaggs

Collegiate 4-H is a student-run community service organization. Collegiate 4-H works on campus and in the community, region, and nation. Projects include sponsoring the Collegiate 4-H scholarship, donation programs, and youth education.

Originally considered as a fraternity for agricultural Extension scholarship members, Collegiate 4-H was first known as Delta Sigma Alpha. Collegiate 4-H began in 1916 on the Oklahoma A&M College campus. The Collegiate 4-H Club was the first in the nation and is the longest-running student organization on campus. Today, the Oklahoma State University chapter of Collegiate 4-H is an organization for any student who is interested in community service, 4-H youth development, leadership, and making great friends.

OSU Collegiate 4-H participates in events including America's Greatest Homecoming Celebration. Each year, the club participates in the sweepstakes competition by competing in events such as the Harvest Carnival, Chili Cook-Off, Lawn Sign Competition, and the Parade Float Competition. Collegiate 4-H also participates in community service projects such as The Big Event.

Off campus, Collegiate 4-H members volunteer as judges for county and state speech contests. They also assist behind the scenes at State 4-H Roundup, as well as sponsor the video and commercial contest at Roundup. Collegiate 4-H members also teach workshops such as parliamentary procedure to local clubs.

Oklahoma Collegiate 4-H participates in various regional and national activities, including regional and national conferences, and has at least one member from the Oklahoma State University club on the regional and/or national officer team. Recently, Oklahoma 4-H's video and commercial contest was converted into a workshop, as well as a competition, at the National Collegiate 4-H Conference.

Collegiate 4-H is working toward being not only a club for students in the College of Agricultural Sciences and Natural Resources, but also for all students on the OSU campus.

Oklahoma 4-H Foundation...
A Growing Legacy

Contributed by Dr. Jim Rutledge

The Oklahoma 4-H Foundation was incorporated on January 25, 1962, in order to receive a generous estate gift from the late R. D. Farmer. The Oklahoma City businessman met Ira Hollar and a group of outstanding 4-H members on their way home from the National 4-H Congress by train. It was then that R. D. Farmer decided to make 4-H the final beneficiary of his estate. Now that his heirs have died, the 4-H Foundation is the sole beneficiary of his trust. This estate gift was finally realized in the summer of 2007, but in the meantime his actions caused the creation of the 4-H Foundation, which has benefited thousands of 4-H members. The initial trustees of the Foundation were Oklahoma A&M President Oliver Willham, State 4-H Leader Ira Hollar, and J. L. Sanderson. For more than ten years, the Foundation had no other directors or paid staff.

In 1974, State 4-H Program Leader Pete Williams and 4-H Specialist Ray Parker completed a process to appoint the first board of directors. An initial meeting of the new board was held on May 22, 1975, at which time new by-laws were approved and the following officers were elected: Ed Synar, president; Pete Williams, vice president; Ray Parker, secretary; and James Denneny, treasurer.

Mr. Parker served as secretary for many years, but most of the work was managing the money that came in from a handful of farm organizations and companies to support various statewide 4-H competitions. Mr. Parker says that the company of Shawnee Milling is probably the longest continuous donor to the 4-H Foundation since its gifts were already coming in when he took over as secretary.

In 1979, Pete Williams gained support from the Foundation Board and Extension Director J. C. Evans to create the first paid position for an executive director for the 4-H Foundation. Retta Miller was the first paid director, and she worked from December 1979 until the fall of 1982. Over the years, a number of people have actually held the title of executive director for the 4-H Foundation, including Miller: Leon Moon, 1982 to 1987; Ray Sharp, 1987 to 1989; Roger Moore, 1989 to 1996; David Sorrell, 1998 to 2004; Jim Rutledge, 2005 to 2007; and Cathy Shuffield, 2008 to present. From 1996 to 1998, board members Glenna Ott and Mary Sue Sanders helped keep the Foundation alive with regular trips to campus. From the beginning to now, the 4-H Foundation executive director position has been jointly funded by the Extension Service and the 4-H Foundation, with the Foundation maintaining its office in the state 4-H headquarters at the Oklahoma State University campus.

From the start, the focus of the Foundation was on promoting excellence among 4-H members. The earliest programs included incentives and awards as well as scholarships for members who excelled at record books or statewide contests. Over the years, support for college scholarships has grown significantly, as have special scholarship and program endowments. During the late seventies and eighties, the Foundation increased the number of trustees, and with the help of the expanded board and early directors, they increased the number and size of gifts through programs like the Order of the Clover and the Alliance for 4-H, which was a partnership with the National 4-H Council. In the nineties, much of the focus was on helping to promote awareness of the 4-H Foundation through promotional events and the "4-H Store" that was set up at all major 4-H statewide events. At the turn of the century, the focus was on finding alumni and asking for their support, which led to increased numbers of endowments and larger annual contributions to support 4-H. Several new donors helped fund 4-H Enhancement Grants, which now provide support for up to fifty club or county groups annually. With the help of board member Larry Derryberry, the Circle of Champions became an important source of unrestricted support. The annual Clover Classic Golf Tournament was also instituted with strong support from the board, including Jerry Kiefer, Gwen Shaw, and Roy Lee Lindsey Jr. The past few years have also included a focus on the organization and management of the Foundation, which was recognized by the IRS as an independent 501(c)(3) in 2008.

Over the years, the Foundation has grown in size and importance to the Oklahoma 4-H Program. Records show that in 1976, the Foundation had assets of $13,944. In 2007, those assets reached an all-time high of 5.3 million dollars. In 2009, annual revenue exceeded one million dollars. The Foundation that was created to allow a businessman to leave a legacy has itself grown into a 4-H legacy. Thousands of 4-H members have participated in events and hundreds have attended college with Foundation support. Past and present donors are commended for their foresight in providing support through the Oklahoma 4-H Foundation.

Oklahoma 4-H: Key Club

Since 1950, the Oklahoma 4-H Key Club has been recognizing the accomplishments of Oklahoma youth in the 4-H program. The club seeks to honor those 4-H members who possess leadership, loyalty, and a sense of responsibility to the total 4-H program.

Membership is an honor extended only to top 4-H members. Membership was once limited to only the top 1 percent of the 4-H youth in the state. Over time, the selection process has been modified to offer the award to top 4-H members who meet established criteria.

Once 4-H'ers become Key Club members, they accept the obligation to support the 4-H organization throughout their lives, both financially and with their time. Key Club members also strive to stay updated with Oklahoma 4-H and the Oklahoma 4-H Key Club.

The key-shaped pin is awarded to new members each year during State 4-H Roundup during a special luncheon and ceremony.

Oklahoma 4-H: National Congress

Contributed by Charles Cox and Tracy Branch-Beck

A train ride to Chicago or a flight to Atlanta are among the memories of generations of Oklahoma 4-H members as they reflect on their participation in National 4-H Congress.

In 1919, a state leader from Tennessee organized a tour of Chicago for one hundred young men and women from around the country who had made a trip to the International Livestock Exposition at the Union Stockyards in Chicago. The tour quickly became an annual event. Youth met to exchange ideas and receive recognition for individual accomplishments and community service. The number grew, and by 1922, this annual event was renamed the National Boys' and Girls' Club Exposition. The meeting in 1922 is officially considered the first National 4-H Youth Congress.

During the seventy-two years in which National 4-H Congress was held in Chicago, over 100,000 delegates, volunteers, partners, and Extension staff members had the opportunity to participate. Congress was sponsored by the National 4-H Council, the Cooperative Extension Service, the United States Department of Agriculture, and numerous industry partners. Youth gathered from all fifty states, the District of Columbia, Puerto Rico, the Virgin Islands, Guam, American Samoa, Micronesia, and the Northern Marina Islands. Getting to attend Congress at the splendid Chicago Hilton was a goal which motivated many 4-H youth. Youth, including those from Oklahoma, were selected to participate based upon their state 4-H record books. While at Congress, they competed to be named national project winners and for the coveted presidential tray.

The 72nd National 4-H Youth Congress, "Celebrating Traditions, Creating Visions," was the last one held in Chicago and was expected to be the last Congress ever held after National 4-H Council and some of the partner sponsors discontinued their support. With the loss of a national awards program, states began to select delegates based upon other criteria in 1994.

When it appeared that Congress would end in 1993, a group of state 4-H program leaders met in San Antonio to plan a temporary event and rallied the support needed to continue Congress. All states and U.S. territories were invited to send delegates to the 73rd Congress held in Orlando, Florida, in 1994. A number of states did not participate that first year but began returning in succeeding years. After Orlando, Congress moved to Memphis, Tennessee, in 1995, and then to Atlanta, Georgia, in 1998, where it continues to be held at this time.

State leaders Jim Rutledge and Charles Cox have both served on the National 4-H Congress board of directors and were involved in its transition to the format that is used today.

Oklahoma 4-H: National Congress

OKLAHOMA
County Histories
4-H

Adair County

Ask anyone in Adair County about 4-H and they can tell you exactly what it means to them because it has been a long family tradition for many, many years. Of course, the livestock projects are still big here, but the trend is extending into other areas, including science and technology.

One of the earliest pictures (at right) is one of a district home economist coming to the Adair County Fair. Notice that she is driving a coupe and the wheel cover indicates the advertisement of the Muskogee State Fair; the year is 1935.

Canning clubs started with using mainly vegetables, but as yields in fruit production increased, many 4-H'ers branched into canning fruits. Adair County 4-H'er Sandy Garrett (on the right) is shown here canning strawberries. *Photo courtesy of the Adair County Extension office.*

In 1960, there were thirty-two clubs located throughout the county. Many were named for places or people in the area and ranged from Chalk Bluff to Mulberry Hollow and from Wauhillau to Clearfork. Today, there are still fifteen clubs in existence in Adair County. We have the largest traditional 4-H population in the Northeast District and the second largest in the state of Oklahoma.

Adair County also boasts many great leaders throughout the 4-H centennial years. Many have become lawyers, doctors, teachers, military personnel, managers of businesses, and overall good citizens. We would like to highlight a couple for you. First is Larry Adair. Adair served in the Oklahoma House of Representatives for several years with the last few as Speaker of the House. Adair still resides in Stilwell with his wife Jan and

is the president of Arvest Bank, Stilwell. Another great leader who began her career in Adair County 4-H is Oklahoma State School Superintendent Sandy Garrett. Garrett is pictured in the photo at left and is on the right.

Adair County is very proud of its rich 4-H history and hopes to continue in the same trend for the bicentennial.

In the early days of 4-H, automobiles were scarce, which left educators to transport the members in wagons pulled by horses. *Photo courtesy of the Adair County Extension office.*

Alfalfa County

Contributed by Tommy Puffinbarger

"Alfalfa County has been spoken of as the Garden of Eden" (*The Republican*, 1928). It has produced fine crops and outstanding individuals.

Boys' and girls' clubs were started as early as 1912 and the county pig club was organized in 1920 with Harold Cross, from Jet, as president. Club work was exhibited in local fairs in 1920. A new county fairgrounds was built in 1927 and Miss Lila State, from Lambert, was crowned the first Miss Alfalfa County. Boys' and girls' club work was changed to Alfalfa County 4-H in 1932.

LaVona Thorndyke of Lamont, Oklahoma, was named the National Achievement Champion at the 26th National 4-H Congress in Chicago in 1947. *Photo courtesy of the Alfalfa County Extension office.*

Former state 4-H ambassadors: Janea Butler, 2004; Shea Masqueiler, 2003; and Joni Puffinbarger, 2001. *Photo courtesy of the Alfalfa County Extension office.*

Alfalfa County and Oklahoma 4-H were well represented from 1937 to 1938 with Amoritas's Bob Morford presiding as state 4-H president. Siblings Wayne and LaVona (Thorndyke) Roush were both national 4-H winners; Wayne in 1941, receiving the Moses trophy, and LaVona in Achievement in 1947.

County agents welcomed the start of the 4-H Adult Leaders organization in 1954 with Mrs. Jay Hertzler, from Aline, as president. The County 4-H Junior Leaders organization was formed in 1966 and many of its original members were national winners, including Ted Weber, from Carmen; Becky Bloyd, from Aline; Gwen and Dixie Shaw, from Burlington; and John and Susan Roush, from Cherokee. Dixie Shaw, John Roush, and Gwen Shaw were inducted into the state 4-H Hall of Fame in 1970, 1972, and 1975, respectively.

In 1971, the Alfalfa County Hall of Fame was started and inducted Loretta Puffinbarger, from Cherokee, and Larry Campbell, from Helena-Goltry, as its first members. Alfalfa County has fifty-eight state Key Club recipients and has produced four state 4-H ambassadors: Ryan Jenlink, from Jet; Joni Puffinbarger, from Burlington; Shea Masqueiler, from Cherokee; and Janea Butler, from Burlington.

"Now, you and I know that the boys' and girls' club work represents one of the most important lines of agricultural activity in the United States today" (Meredith, 1920).

A group of 4-H youth dedicating an Oklahoma Red Bud tree in 1938. L-R: Velda Swiggert, J. Lawrence Hague, Lester Whitney, Geraldine Riley, Milton Hague, and Guy Swiggert. *Photo courtesy of the Alfalfa County Extension office.*

Atoka County

Contributed by Lynne Beam, Thelma Cooper Brady, and Lee Harbin

Atoka County 4-H currently has 416 4-H members enrolled in sixteen clubs, with twenty-four adults volunteering their time to serve in leadership roles as they strive to provide opportunities to develop important life skills in our youth. Members have diverse interests in projects from photography to livestock, clothing and textiles, and food preservation.

Thelma Cooper Brady, an alumna of Lane 4-H, recalls that her most memorable 4-H year was 1950, which she calls "The Year of the Peanut Butter Cookies."

"My team member and I made lots and lots of peanut butter cookies practicing for competitions. When the home demonstration club met, we were there to practice our demonstration. We won Grand Championship at the county level and that summer went to Oklahoma A&M in Stillwater for 4-H Roundup, where we placed in the blue ribbon group. I'm sure we did more than give our demonstration, but peanut butter cookies are my main memory. After 4-H, I have used the skills learned in cooking, public speaking, and sewing many times, but as memorable as the cookie-making demonstration was, it was many years before I ate peanut butter cookies!"

When asked by daughter Lori Boehme and granddaughter McKenzi Boehme what he remembered most about his 4-H experience, Lee Harbin, alumnus of Caney 4-H Club, replied: "The summer of 1968, at the age of seventeen, I participated in the Oklahoma 4-H Delegation's twenty-eight-day trip to Europe. By air, rail, and bus we toured eight countries: Czechoslovakia, Switzerland, Norway, Belgium, Holland, Sweden, France, and Germany. The purpose of the tour was to learn different cultures and farming techniques of the European countries. We toured dairies, farms, wineries, cheese factories, churches, cathedrals, and attended a ballet and an opera. This 4-H trip was a great opportunity and learning experience."

In 1951, Thelma Cooper Brady of Lane 4-H was busy sewing her garment for the spring 4-H dress revue. *Photo courtesy of the Atoka County Extension office.*

Top right: In the summer of 1950, Claude Collier Chevrolet & Buick Company in Atoka, Oklahoma, presented the Atoka County 4-H and the Atoka FFA each with a new pickup as the new models came out. *Photo courtesy of the Atoka County Extension office.*

Practice makes perfect as Thelma Cooper Brady of Lane 4-H makes a batch of home-canned vegetables. *Photo courtesy of the Atoka County Extension office.*

Bottom right: In the summer of 1968, Oklahoma 4-H took a delegation on a twenty-eight-day trip to Europe. By air, rail, and bus they toured eight countries. The trip's purpose was to learn the different farming techniques and cultures of the European countries. *Photo courtesy of the Atoka County Extension office.*

Atoka County

Beaver County

Contributed by Sandra Cooper and Rachel Hayes

The 4-H boys and girls loaded up in a large cattle truck to go to Goodwell, Oklahoma, to go to training school, where they would learn how to help farm labor and production shortages. *Photo courtesy of the Beaver County Extension office.*

Extension work was started in Beaver County in 1915 by J. T. Newson and his wife. In 1922, Ruth Randal began adult clubs, and by the end of the year there were 109 girls enrolled in junior work. By 1923, there were only girls' clubs, and they were only enrolled in work areas they desired, such as sewing, canning, etc. By November 1, 1926, there were 447 4-H members enrolled in Beaver County. During the start of the Depression in 1932, Beaver County 4-H girls made 1,963 articles of clothing and canned a total of 3,844 quarts of food. It was hard to keep the boys interested in 4-H club work during the Depression because crops were not growing and with no feed, livestock had to be sold or slaughtered. In 1935, six 4-H members attended State 4-H Roundup, it being the first time anyone from Beaver County attended. As a result of the National 4-H Victory Program, eighty-five 4-H boys and girls loaded up in the back of a large cattle truck and went to Goodwell, Oklahoma, to training school to learn how to help farm labor and production shortages. In 1957, the Beaver County Junior Leaders organization, Teen Leaders, planned and conducted two different exchange trips with Clay County,

Kent Perkins demonstrates how to "Beat the Heat with Swine." *Photo courtesy of the Beaver County Extension office.*

Beaver County

Above: The 1991 Senior and Intermediate Oklahoma State Horse Quiz Bowl champions. L-R: Amber Stanley, Dr. David Freeman (OSU horse specialist), Dirk Stanley, Richie Dorman, Rissie Heglin, Eric Duncan, Sara Manuel, Dan Wall (OSU grad student), Kodel Heglin, Dusty Matthews, Dr. Don Topliff (OSU animal science), and Tracy Devenport. *Photo courtesy of the Beaver County Extension office.*

Right: Youth at 4-H summer camp in 1961 learn how to make jelly. *Photo courtesy of the Beaver County Extension office.*

Beaver County

A group photo of the Beaver County 4-H award winners at the 4-H Achievement Banquet, January 9, 1956. *Photo courtesy of the Beaver County Extension office.*

Missouri, and Hardin County, Kentucky. Everyone enjoyed the trips and getting to know other states' 4-H programs.

Throughout the years, Beaver County has had several state officers: Jim Loepp, 1964 state 4-H president; Rodney Albert, 1983 state 4-H Northwest District vice president; Lindsay Sherrer, 1994 state 4-H Northwest District vice president; and Melissa Barth, 2007 state 4-H reporter. Beaver County 4-H has been strong throughout the years and continues to uphold its traditions. We are looking forward to another one hundred years of great memories and opportunities.

Beaver County 4-H at the State 4-H Roundup in 1961. *Photo courtesy of the Beaver County Extension office.*

Beckham County

Contributed by Opal Nelson and Herbert Hartman

Oklahoma 4-H has always been a very important part of the fabric of Beckham County. In the early 1930s, young men were instructed in animal judging and animal husbandry and young ladies were taught skills in canning, sewing, and quilting. These "hands-on" lessons were taught by many parents, volunteers, and older students. With the help of these individuals, along with the 4-H agents in the county, 4-H membership increased with a tremendous number of students becoming involved. Under the helpful guidance of men like P. G. Scruggs, Robert Reeder, Wallace Smith, and Herman Seymour, Beckham County 4-H thrived from 1935 through the mid-1980s.

The 1947 Beckham County group at Roundup in Stillwater. *Photo provided by Herbert Hartman.*

Beckham County 4-H group representatives in Stillwater, Oklahoma, for dress revue in 1946, in front of Old Central. *Photo provided by Herbert Hartman.*

Beckham County

The 1947 state livestock judging team champions. L-R: a state representative (unknown), W. O. Baker, Henry Howard Hanni, Herbert Hartman, and Robert Reeder, the Beckham County agent. *Photo provided by Herbert Hartman.*

In the thirties and forties, school officials saw the need for leadership programs that the Oklahoma 4-H clubs could provide. Many rural schools like New Liberty, Doxy, Highway, Delhi, Carter, Retrop, Erick, Sweetwater, and Hext started 4-H clubs in Beckham County. Each of these clubs grew in membership with the help of parents and volunteers; from that point on, Beckham County 4-H was established as a true contender on the state and national level.

Beckham County 4-H has been extremely successful on the state and national level with many winners, including Paul Mackey, 1980–1981 state 4-H president, and W. O. Baker, Henry Hanni, and Herbert Hartman, 1947 state 4-H livestock judging team champions and champions at American Royal.

Beckham County is continuing to grow, with four 4-H clubs with over 170 members and thirteen volunteer leaders.

Mike Nelson, 1974 Oklahoma Outstanding 4-H Member. Also pictured are Mike's parents, Mr. and Mrs. Bruce Nelson. *Photo provided by Opal Nelson.*

Blaine County

Contributed by Michelle and Sarah Helm

Blaine County covers 583,040 acres located in the northwest part of Oklahoma. The population is 11,976 people.

As we page through history, we see a common thread that holds the people of Blaine County together: people serving the community by informing and educating. At first, 4-H officials began activities by doing home demonstrations. These activities were a vital part of rebuilding the quality of life during and after the Great Depression.

In 1928, Ola Armstrong began her career as a home demonstration agent; she had 158 participants. That same year brought the county's first 4-H Club Rally, with each club participating in the parade down Main Street of Watonga. Today, the tradition is to have a 4-H float in the Watonga Cheese Festival Parade.

During the 1940s and 1950s, Blaine County began an achievement program. There were 323 boys and 284 girls enrolled in 4-H clubs. These were hard times in rural Oklahoma with the sugar shortage and war rations. The Extension office served the community by setting up war bond sales and Feed and Seed loans. Mrs. A. M. Barrows sums up a sign of the times in her quote taken from the Narrative Annual Report of Extension Work, 1941: "We pause to give thanks to God, that we are in America today and can have these bright prospects of the future."

In 1963, Blaine County Livestock Club promoted the

Blaine County 4-H members in Stillwater for the 1931 Roundup. *Photo courtesy of the Blaine County Extension office.*

A 4-H rally was held on Main Street at Watonga, the county seat, in 1937. *Photo courtesy of the Blaine County Extension office.*

Blaine County

The 4-H youth participate in a writing workshop at the 2008 camp. *Photo courtesy of the Blaine County Extension office.*

first Bonus Auction Program, raising $2,000. It is a time-honored tradition that still exists today. This year's sale raised approximately $76,000 with all children who exhibited livestock participating.

Many families work hard all year to produce award-winning wheat, livestock, and other home projects, taking top honors at district and state fairs. The shooting sports program is very competitive at the national level. Nonetheless, our county produces some of the best speakers and musicians and vocalists, with many participating in the district Share the Fun.

Leaders and members of 4-H attend a training meeting for canning and food preservation in 1941, presented by Martha McPheter, Extension staff. *Photo courtesy of the Blaine County Extension office.*

Bryan County

Contributed by Maddi Shires

Bryan County's rich 4-H history is evident everywhere. Some Bryan County 4-H families go back four generations. County schools and homes have display cases full of 4-H trophies, ribbons, and plaques.

Bryan County's Extension history began in 1908 with J. J. Ross as the first county agent. Visitors to the Bryan County Extension office immediately see its rich history by glancing at the Hall of Fame. With the exception of one year, Bryan County has had at least one member inducted into the Hall of Fame since 1971. Bryan County has had 143 Key Club award recipients, many district officers, a few state officers, and three state ambassadors.

Many former members have had successful careers as doctors, lawyers, teachers, public officials, and financial experts. All of them credit 4-H with helping them become who they are today.

Youth display their finished products from the Bread-making Day Camp in 2006. *Photo courtesy of Robert Bourne, Extension educator.*

Youth participate in the county Share the Fun Talent Show, 2009. L-R: Clay Shires, Rachel Childers, Jake Shires, and Teigan Munson. *Photo courtesy of Robert Bourne, Extension educator*

Youth participate in the 2008 Roundup. L-R: Nikki Schuth, Leslie Carter, Zack Childers, Ethan Schuth, Shacole Smith, Lynsi Bourne, and Maddi Shires. *Photo courtesy of Robert Bourne, Extension educator.*

Katy Ann (Boyd) Glover remembers the Bryan County officer team's 1999 annual retreat to Grapevine, Texas. Although leaving in plenty of time for 7 p.m. dinner reservations, they were delayed in miles of traffic by a crane that had fallen off a truck. They passed the time by rolling down the windows and talking to people. They were able to help others merging into traffic and get information about what routes to take. They arrived at 9:30 p.m. that night, missing dinner reservations by hours, but made new friends and learned that they could take any situation, make it fun, and be leaders, which was the whole purpose of the trip.

Currently, Bryan County has seven clubs and 556 members enrolled. Robert Bourne, Bryan County Extension 4-H educator, along with numerous volunteer leaders, is working to promote a large variety of 4-H activities. Bryan County has poultry, horse, horticulture, and shooting sports specialty clubs. Many members still take interest in traditional activities such as showing and caring for animals, but many youth today have other interests. Judging clubs, Share the Fun, and specialty clubs have some of the greatest interest in Bryan County today.

Caddo County

Contributed by Kevina Green and David Nowlin

Caddo County is primarily a rural county, and continues to have one of the largest 4-H enrollments in the state of Oklahoma with twenty-two active 4-H clubs. The membership is as culturally rich and diverse as the topography that covers all 1,290 square miles of the county. The county seat, Anadarko, remains unique in all Oklahoma cities because the majority of its citizens are Native Americans. Several tribes, such as the Kiowa, Delaware, Apache, Caddo, and Comanche, just to name a few, have a strong presence in Caddo County.

In addition to having a rich tribal history, Caddo County has a strong rural and agricultural history. Caddo County is number one in the state for peanut production and continues to be one of the largest cow/calf and stocker operation counties. Caddo

For many 4-H members, community service is a big part of their 4-H life. These Caddo County members are putting together care packages. *Photo courtesy of the Caddo County Extension office.*

Shooting sports is a very popular project area for Oklahoma 4-H members. *Photo courtesy of the Caddo County Extension office.*

Caddo County

Even though Oklahoma 4-H is branching into more urban and scientific project areas, for many active members, livestock production and exhibition are still popular. *Photo courtesy of the Caddo County Extension office.*

The beginning of 4-H in Caddo County is much like many of the counties in the great state of Oklahoma. Caddo County was one of the first counties to have U.S.D.A. Farmer's Cooperative demonstration workers. One of the first was Mr. F. F. Ferguson in 1909, who worked in Caddo, Canadian, and Grady counties. One of the first home demonstration workers was Miss B. R. Hudson, who was hired in 1912. As is the case for most counties, the beginnings of 4-H started with corn and canning clubs.

Things really got rolling in 1914 with the Smith-Lever Act creating the Agricultural Extension Service. The "county agents" would work with the youth, teaching them new and innovative skills in agriculture and homemaking with the hopes that the young people would take these new research-based ideas home where the families would adopt the new practices. In the beginning, those Extension County was founded in 1890. Caddo County consists of several small rural communities including Binger, Boone, Carnegie, Cement, Cyril, Eakley, Ft. Cobb, Gracemont, Hinton, Hydro, Lookeba, and Sickles.

employees who assisted with 4-H and Youth Development were called county agent assistants. The first in Caddo County was Miss Elizabeth Givens, who was hired in 1938.

Since the beginning, there have been thirty-two 4-H/Youth Development Extension educators. The current one is Ms. Kevina Green. Ms. Green has been in the position since 2005. Over the years, the 4-H clubs of Caddo County have seen many changes, as have all of the clubs in Oklahoma.

Presently, there are twenty-two community clubs, a shooting sports project club, a horse project club, and a teen leader organization with over 700 members enrolled. Caddo County is fortunate to have wonderful school support from several of the small communities making up Caddo County. This county can boast over 300 speeches and demonstrations given at the yearly contests. Caddo County continues to have the highest number of youth attending the Southwest District Youth Action Conference. Additionally, Caddo County is home to four state 4-H presidents.

Caddo County 4-H is rich in tradition! With these strong roots, the outlook for Caddo County 4-H looks great.

Canadian County

In the Canadian County Narrative Annual Report of Extension Work, 1927, Home Demonstration Agent Irene Hanna states: "When I began in Canadian County in November, there was one active junior 4-H club of twenty-eight members and a few boys and girls over the county carrying on project work in 4-H demonstrations work. The county organization began by my notifying the seventy-seven rural teachers of my presence here."

Fred E. Percy, county agent, reports that a centralized system of organization was followed in carrying out the junior 4-H program. At the time, eighteen centers were selected and the adjoining schools grouped in the centers according to accessibility to the center.

In 1936, Miss Harvey Thompson, home demonstration agent, reports having sixteen 4-H clubs with a total membership of 500. She reported, "The 4-H clubs met continuously all the year not missing any meetings because it was summer or harvest, etc." To promote interest in 4-H club work, a community night was held with each club being allowed to use any ideas of its own to make the program most effective in its particular community. "These programs did great good in awakening interest among parents in seeing that their 4-H boys and girls were given encouragement at home in doing the things they learned in 4-H club, and since it takes home support to make 4-H club work most effective, we are very pleased over the results gained through these programs," the report continued.

Project areas in which end-of-the-year prizes were awarded are listed as garden and canning, food preparation, home improvement, and sewing for the girls, and livestock, crops, dairy, and poultry for the boys.

In 1942, Mr. Lee Phillips was the county agent and there were 401 boys and girls enrolled. Mr. Phillips reported that 4-H "cooperated with the teachers by grading the seventh- and eighth-grade boys and girls so that they could be given homemaking or agriculture credit for their 4-H club work." Fifty-seven team demonstrations were given by 4-H club girls. Individual premiums for the county fair were paid in War Savings Stamps to ninety-eight 4-H and FFA members of the county. That year, 363 exhibits were made, including 157 girls' exhibits in home improvement, clothing, canning, and food preparation; 158 in livestock; and fifty-four in crops and horticulture.

The 1943 Canadian County dress revue, best of class.
Photo submitted by Debbie Baker.

Canadian County

The 1995 Canadian County delegation to State 4-H Roundup. *Photo submitted by Debbie Baker.*

In 1944, Canadian County had 393 members in seventeen clubs and Canadian County held their first annual achievement banquet. Patricia Meyers, age twelve from Calumet, stated about her clothing 4-H work: "I also entered my dress (in the county fair) that had been made out of flour sacks and it cost fifteen cents. I am wearing it to school now and I like it very much. The cost was the tape and thread."

In 1953, Mr. Riley Tarver was the county agent, and there were 303 girls and 301 boys enrolled in the 4-H program. For National 4-H Club Week, March 7 through 14, Sandra Chiles prepared a white streamer, lettered in green, which was stretched across South Rock Island Street, highways 66 and 81 through El Reno. The county contest for 4-H boys and girls was held in April and approximately 1,000 attended; contests included timely topics, demonstration, dress revue, and health contests.

In 1968, Margaret E. Fitch and June Pearson Cash were the Extension home economists for the 4-H program. Two 4-H members, Jayne Reuter and Helene Halacka, conducted six charm school sessions for which sixty girls received certificates in the graduation service. There were 436 entries in the county fair, not including livestock. Monthly county recreation was held at Banner School, where folk games and relays were enjoyed by approximately eighty members.

In 1972, Canadian County had 2,559 4-H members enrolled in twenty 4-H clubs and Betty Fleck was the 4-H Extension agent. Trips taken by 4-H'ers included National 4-H Congress (Chicago), National 4-H Dairy Conference (Madison, Wisconsin), National

Safety Congress (Chicago), National Citizenship Short Course (Washington, D.C.), American Royal 4-H Educational Conference (Kansas City), and State 4-H Club Congress (Oklahoma City).

In 1976, 4-H record book pins were given to 177 members and special awards were given for demonstrations in the areas of electricity, commodities, breads, dairy, and peanuts. The Oklahoma Cowbelles, the women's beef cattle organization, sponsored an award for beef selecting, preparing, and serving. During the seventies, 4-H members from Canadian County and Custer County gathered photos from Oklahoma to make a collage and it was taken to the national 4-H office in Maryland and displayed on the wall outside the Oklahoma Room until 2003.

In 1986, Susan Meitl and Jeff Lorah were the 4-H Extension agents for Canadian County and 548 members participated in twenty-two clubs. The 4-H program also sponsored a school enrichment program, which provided teachers throughout Canadian County with educational materials such as chick embryo science, entomology, meteorology, nutrition, and plant science. The 4-H program conducted eighty-eight school enrichment programs with 2,143 students reached.

In 1997, approximately five acres of the Kirkpatrick property was dedicated to Canadian County 4-H for a project area. The Kirkpatrick family donated funds to build a barn to educate urban youth and families about agriculture. The 4-H program was able to greatly expand its urban education and outreach, and within only a few years it became a vital part of Canadian County 4-H, where more than 1,500 visitors participate in over seventy activities each year.

In 2009, Canadian County had twenty-one clubs consisting of 586 members and Tom Manske (1993–present) and Susan Meitl were the 4-H Extension agents. A scholarship in memory of 4-H Extension Agent Susan Meitl was established. A mural was painted by 4-H member Kenna Maria Baker to celebrate the centennial of Oklahoma 4-H, which was displayed at the State Capitol, then at the state 4-H office for one year. Also in 2009, Sam Durbin became the first Canadian County 4-H member to be elected as the state 4-H president. The four most popular project areas include livestock, photography, shooting sports, and the horse project.

Carter County

The history of 4-H in Carter County reaches back to June 14, 1906, with the appointment of D. A. Caldwell as the first Extension agent. The very first mention of 4-H is in 1910, when Floyd Gayer of Ardmore won an all-expense-paid trip to Washington, D.C. The trip was offered by Senator Thomas P. Gore to the Oklahoma boy who produced the largest yield of corn. Floyd had a yield of more than ninety-five bushels per acre.

Reading through reports made by Extension agents indicated that early clubs were called Boys' Clubs and Girls' Clubs. During the early years, projects for boys and girls were separate and contests were separate. The project exception was dress revue. It seems that it has always been important for youth to learn how to dress properly for the occasion. It was not until the early seventies that appropriate dress, as it was then called, was for boys and girls together. Even then, awards were given to best overall girl and boy in both junior and senior divisions.

Records from the twenties to mid-thirties show a picture of an organization that was well received by all schools in the county. Many of the "coaches" who led the groups were older 4-H members who had seen the benefits of what they had learned and wanted to insure that this opportunity was offered to other youth in their communities. One such girl was Robbie Brown of Graham as reported by Extension Home Demonstration Agent Minnie B. Church. "Robbie began doing club work eight years

Carter County

ago when her parents lived at the Brown school house. The father moved to California but did not like it so came back to Carter County at Graham where Robbie entered school and found 4-H club work was being neglected. She organized a club of 43 members" (*Daily Ardmoreite*, 1922).

The most popular 4-H event now is the junior livestock show. The show was first started on November 19 through 21, 1936. The 1978 show was dedicated to Ikey Johnston and Earl Wallace, who were the first two exhibitors to show calves in the 1936 junior livestock show.

In the thirties and forties, camp was held in the Arbuckle Mountains. Later in the sixties, camp was held at Lake Murray. In 2009, Carter County was in the Arbuckle Mountains (a different location) once again. It was three days sleeping indoors at ninety dollars compared to a whole week sleeping outdoors for fifty cents and vegetables from home.

Cherokee County

These two 4-H'ers pause for moment while working in their county office to take a picture. *Photo courtesy of the Cherokee County Extension office.*

Cherokee County 4-H has a rich history with several families having three or four generations of membership, leadership, opportunities, and friendships, now with Carl Wallace as Extension educator.

Four generations of Yevonne Loftin's family have been in 4-H. In the early thirties, her mother, Lois Garoutte Collier, was a member of Grandview 4-H.

"Traveling back then was by horse and wagon, but the 4-H'ers got to ride a school bus to Wagoner for a fashion revue. Mom won a blue ribbon." In the fifties and sixties, Loftin enjoyed 4-H rodeos. She was fortunate to be "the Roundup Club Queen [and] to travel the state promoting 4-H at fairs and horse events." She was a 4-H adult leader for thirty years, and both her daughters, Jamie Loftin Cole and Anita Loftin Jones, were county Hall of Fame winners and state and regional winners in entomology, with Jamie a national winner. Both traveled to National 4-H Congress in Chicago and to Washington, D.C., for a Citizenship Focus Convention. "Now my grandchildren are experiencing the great rewards of 4-H," Collier said.

Share the Fun was always a big deal at Lost City, thanks to Mr. and Mrs. Danny Bilby, Margie Phillips, Yevonne Loftin, Anita Carey, Pauline Howe, and others, Jamie Loftin Cole said, and dress revue always had fifty to eighty girls participating. She attributes her success in life to Dean Jackson, Rick Rexwinkle, Marty Green, Virginia Price, Eula Mae Miller, Wanda Bolding,

Cherokee County

David Campbell, and Larry Sams with the district office for guiding her on her 4-H journey.

Bob Kennedy, county agent in Cherokee County from 1960 to 1975, recalls a Government Day Program.

"Our 4-H kids elected their own county and state government officials and studied government," he said. "Senator Fred Harris, the state Extension director, and other dignitaries attended our final program." He came here for rural development from Muskogee County. "We had about thirty clubs, a strong livestock program, and we took large delegations to Roundup."

Attorney Tina Glory Jordan joined 4-H in 1965. She earned Hall of Fame with her siblings, Trail, Blaise, and MarDon, and her children, Drew Jordan and Cray Jordan King. Three grandchildren now are in 4-H. "Bud Curry was the agent when I was nine. He took us to Tulsa on a judging trip and I won a twenty-dollar gift certificate. We didn't go to Tulsa very often in those days but mom loaded me up the next week and took me to Tulsa to spend it."

Mother Glory was a 4-H'er, a forty-year volunteer, and a Hulbert School teacher. "We found her first entry into the fair when she was nine, a handkerchief with her name and first-year 4-H sewn on it." Glory also recalls winning her first grand champion in the sale barn that Charlie Kirk and Carl Greenhaw owned. "You can see an air of confidence in children who are in 4-H," Glory said. "It keeps a child busy and out of trouble, and brings a family closer together. [It] positions us to be more in life, strive in business, and succeed."

These young members are using a fun way to learn about health and nutrition with fruit. *Photo courtesy of the Cherokee County Extension office.*

Marty Green, the 4-H agent in Cherokee County from 1984 to 1995, said they raised their own funds to take two exchange trips to Alligan County, Michigan, and Franklin, Pennsylvania. "We also took an educational tour to California. When I see those kids today, they still talk about those trips." District speech contests started here, along with tri-county day camps, shooting sports, and the state broiler show they hosted for years, he added. "We had one of the first nationally certified instructors in Oklahoma for shooting sports, Vernon Martin."

Basket weaving has become a fun and unique project area for Oklahoma 4-H members. *Photo courtesy of the Cherokee County Extension office.*

Choctaw County

There are five active 4-H clubs in Choctaw County. They include the Ft. Towson, Hugo, Soper, Horse Club, and Boswell. The 4-H educators for the county are Marty Montague and Tracey M. Watts. There are approximately 200 4-H members. Summer workshops held each year are well attended and help with the recruitment of new members.

The program's focus areas range from forestry, land judging, and horse judging to showing livestock. We have 4-H'ers who have gone on to a collegiate level in some of the judging areas as well as those who have become an influence in the decision-making of career choices for some of our youth. The 4-H program has had members become state record book winners and district officers, and a number of others have achieved honors at the district, state, and national levels.

These 4-H members enjoying 4-H camp take a moment to pose for a group photo. *Photo courtesy of the Choctaw County Extension office.*

These 4-H members are participating in 4-H Share the Fun skits. *Photo courtesy of the Choctaw County Extension office.*

Cimarron County

Contributed by Ferrell Ted Smith and Joyce Wells

Cimarron County 4-H has been an important part of the life of the county since the early part of the twentieth century. Many remember when W. E. Baker was the county agent during the 1930s. Others recall Eugene Williams and Ferrell Ted Smith. Some of the home demonstration agents mentioned are Pat Widener Frances, Barbara Bryant, Louise Fairchild, Sandra Turner, Liz McBee, and Marty Albright.

Ted Smith was Cimarron County Extension agent, 4-H agent, and Extension director from July 1958 until his retirement on March 1, 1987. During that time, there was always a full delegation that went to the 4-H Roundup in Stillwater. The Cimarron County Fair was packed with exhibits from every club in the county. They represented the clubs from Boise City, Plainview, Felt, and Keyes. It was also during this time that the junior fat stock show was organized by Ted Smith. The county rally and dress revue was an all-day activity with almost every child in the county participating.

Many of the 4-H members have become outstanding members of their communities. Besides the many farmers and ranchers and business women and men, 4-H members have become nurses, doctors, pharmacists, physical therapists, a district judge, a district attorney, a mayor, county commissioners, and schoolteachers who have trained hundreds of young citizens.

Cimarron County no longer has an Extension office, but Plainview 4-H Club in the Griggs Community is an example of a club that is still striving to "make the best better." Since it was organized in 1923, Plainview is eighty-six years old and still has active members. The other three clubs also have excellent members with devoted leaders.

Cleveland County

Contributed by Brenda Hill and Justin McConaghy

Cleveland County's Extension office opened on November 1, 1909, and H. Garland was our first Extension educator. We are unsure exactly when 4-H clubs were formed in Cleveland County and the earliest photos available were from the 1940s. However, since 4-H began in our county, thousands of youth have experienced the satisfaction of being a 4-H'er.

One of our first Cleveland County Key Club members was Mary Gabrish Offutt who was installed in 1951. Mary was the first of a long line of Key Club members in our county. In 1950, the president of

Sidney Calvert of Cleveland County 4-H with his show dairy heifer and Phil Kidd of First National Bank in Norman, in the 1940s. *Photo courtesy of Sidney Calvert.*

Cleveland County

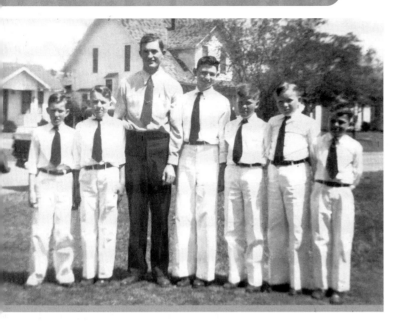

Cleveland County 4-H Educator Denver Patterson and the boys of the Pleasant Valley 4-H Club in the 1940s. *Photo courtesy of Sidney Calvert.*

Cleveland County has also had many outstanding club leaders and volunteers. Lydia Calvert Webb was one of our past club leaders. She is ninety-six years old and is still very active in her HCE (Home and Community Education) group. Her oldest son, Sidney Calvert, is seventy-three and was in Pleasant Valley 4-H in the 1940s and his daughter, Shelly, was in the same club in the 1970s.

Cleveland County is still producing generations of 4-H'ers. We currently have about 400 4-H members, many of which are third- and fourth-generation 4-H families. Today, we still offer traditional 4-H projects such as animal science and family and consumer sciences while also staying on the cutting edge of technology, offering projects like robotics.

Sears, Roebuck and Company, F. B. McConnell, wrote to Mary: "We know that better family life today and better citizenship for the future result from work such as you have done" for her national 4-H home improvement win sponsored by Sears. Other Key Club members from Cleveland County in 1951 were John B. Morrem and Earl W. Morrem. We have continued to have several state Hall of Fame winners and numerous record book winners at both the national and state levels.

Mary Gabrish Offutt, Cleveland County 4-H Key Club inductee in 1951. *Photo courtesy of Mary Offutt.*

Coal County
Contributed by Bette Price and Melba Thomas

Coal County 4-H'ers attend Roundup at Oklahoma State University. L-R: Jeri Rae Hampton, Jaylene Ward, Lana Thomas, and Tammy McCurry. *Photo courtesy of the Coal County Extension office.*

Bette Price says that having worked in the Coal County Courthouse for thirty-two years, she has seen the ups and downs of the 4-H and other Oklahoma Cooperative Extension programs. Funding was the biggest concern, so Jan Montgomery, retired district director for Oklahoma State University Extension, helped us pass a sales tax in Coal County. Our 4-H'ers and Extension homemakers sponsored fundraisers to help fund the office before the county sales tax was passed.

Landon Garcia, a 4-H exhibitor, is pictured with her county fair exhibits in 2006. *Photo courtesy of the Coal County Extension office.*

friends. It has provided endless entertainment while instilling the qualities of self-esteem, perseverance, and integrity. Their experiences are invaluable in providing an educational experience that no other organization provides. My second daughter, grandson, and two granddaughters are Coal County Hall of Fame recipients. Thank you, 4-H, for one hundred years of providing a timeless organization that has remained true to its participants and will continue to provide opportunities for young people to follow their dreams."

In the late 1980s, Price's children, Chris and Jamie Price, were very active in Coal County 4-H. Both held county and district offices during their years in 4-H and were selected as Hall of Fame winners. They helped with all sorts of workshops and camps for the younger 4-H'ers. Because of Chris's 4-H involvement, Mr. Grant Praytor, Extension agent at the time, encouraged Chris to attend OSU Tech at Okmulgee and took him on a college tour. He graduated from OSU Tech at Okmulgee in 1990 and started his own company, which he still operates. Jamie attended the Washington, D.C., Citizenship Focus trip in 1989, which was an experience she shares with her children when studying about our nation's capitol. The 4-H program builds character and self-esteem and provides positive life experiences to the youth and volunteers in Coal County. That is why families continue to be active in 4-H.

Melba Thomas says her involvement in 4-H began forty-one years ago with her oldest daughter. "Since that time, I have watched my four children and seven grandchildren participate in 4-H. Though there have been some changes throughout the years, 4-H has remained constant and a valued part of my family. [It] has provided a venue for working together as a family and for sharing fun with

Landon Garcia won the Share the Fun Junior Individual category with a dance performance. *Photo courtesy of the Coal County Extension office.*

Craig County

Contributed by Tari Lee, Roy Ball, Dotty Daniels, and Rose Mary Johnston

The beginning of 4-H in Craig County has been hard to pinpoint. We know of one member, Albert Minson, who was once honored as the most senior former 4-H member in Craig County at the time. He joined the Boys' and Girls' Demonstration Club at age thirteen, when he was a student at Couch School. He said that it wasn't called "4-H" when he joined in 1918. His first project was growing corn. He said he grew seventy ears of corn, which made one bushel. It usually took 120 ears to make a bushel, but his corn grew real big that year because it was on "new ground." His corn was put on display at the Craig County Fair and was bought for ten dollars by a local real estate agent.

Craig County's first county agent was G. E. Thomas, who worked from 1913 to 1921. The 4-H clubs were based out of the rural schools in the county, where everyone was in 4-H!

In 1943, 4-H members across the state sold war bonds. They helped to purchase thirty B-24 Liberator bombers with the $9 million they raised. They later helped the U.S. War Department fill sacks with the seed pods from milkweed plants, used as filler in U.S. Navy life jackets. The 4-H clubs from Craig County alone filled over 886 sacks with the seed pods!

In the 1950s, the Craig County Fair was held at the Vinita Livestock Sale Barn. The 4-H awards banquet was held in the Vinita Hotel ballroom. Events were primarily held at the Courthouse Annex. Now, the fairgrounds serve as the location for most events.

From the corn and tomato clubs of 1918 to the science and technology of 2009, 4-H has come a long way in Craig County for the development of the youth that serves our communities.

Albert Minson of Vinita was a former member who started his career in 1918. *Photograph courtesy of Rose Mary Johnston, daughter of Albert Minson.*

Top right: These 4-H youth show off their steers at the Craig County Fair, which was held at the Vinita Livestock Sale Barn. *Photo courtesy of the Craig County Extension office.*

Bottom right: Craig County 4-H girls during a 1950s dress revue. *Photo courtesy of the Craig County Extension office.*

Craig County

Craig County

In the 1950s, 4-H camp was held near Catoosa, Oklahoma. *Photo courtesy of the Craig County Extension office.*

Creek County

Since the establishment of their first 4-H club in the early 1900s, Creek County 4-H has strived to meet the needs of youth and adults in rural communities. Since its inception, Creek County 4-H has focused on providing a safe and inviting environment that provides youth the opportunity to "learn by doing" and develop life skills. Throughout the years, Creek County 4-H has maintained this focus and continues to offer learning opportunities in project areas such as agriculture, animal science, citizenship, leadership, and family and consumer science.

In the 1940s, Creek County 4-H maintained this excellence under the direction of 4-H Agent Mr. Jay Hesser. Mr. Hesser reflected upon his time in Creek County as a tremendous opportunity that allowed him to work with thousands of youth in developing various 4-H projects. Mr. Hesser was instrumental in developing poultry contests and livestock judging teams, some of which competed at the national 4-H contest in Chicago, Illinois. A poultry contest founded by Mr. Hesser is still held in Drumright, Oklahoma, each year and serves as testament of the traditions started in 1945.

Today, Creek County 4-H continues to offer educational and service learning opportunities to over 2,150 youth through 4-H project work and school enrichment programs. While Creek County 4-H continues to focus on traditional 4-H project areas, it has also diversified its approach by offering additional projects in the areas of science and technology and shooting sports. The 4-H program continues to be a vital instrument that provides youth with a foundation of learning that will follow them throughout their adult life and develop our young men and women into productive citizens for the future.

Custer County

Contributed by Ron Wright, Radonna Sawatzky, and Taler Sawatzky

The Custer County 4-H Program has always had a strong tradition in Custer County. Many citizens in Custer County were once 4-H members and are continuing that tradition with their children, grandchildren, and great-grandchildren as 4-H members. The program started in February of 1912. The Custer County 4-H Program has seen enrollment as high as 800-plus members. The current 200 members today are as competitive as ever. The program has evolved into many new and innovative projects while still holding true to the traditional 4-H projects that have been the mainstay of the 4-H program for years. Custer County 4-H hosts speech and illustrated presentation contests, fashion revues, impressive dress, job readiness, favorite food show, Share the Fun, 4-H camp, summer workshops, and much more. Record books have also been a tradition and have helped Custer County 4-H members become national and state project winners. Many 4-H members have been state 4-H Hall of Fame winners, Key Club members, state ambassadors, and district and state officers.

Custer County members have been winners in all kinds of competitions, from the local, county, and state levels, to even traveling by railcar to California in the 1930s to compete and win at a national livestock show.

Custer County 4-H members really shine when it comes to community service projects. The Custer County Teen Leaders exemplify true citizenship by giving back to their communities through numerous projects like Relay for Life, Agape Food Bank, Veterans Center projects, nursing home visits, Habitat for Humanity, a children's shelter, Christmas Connection, cards for soldiers, stockings for foster children, Ronald McDonald House, and many others. Custer County 4-H continues the tradition of excellence and looks forward to the next one hundred years.

Delaware County
Contributed by Barbara Denney, Dian Ousley, and Dianna Park

Left: A Delaware County 4-H canning demonstration at the home of Mrs. L. T. Mayfield, near Grove, Oklahoma, in 1929. *Photo courtesy of the Delaware County Extension office.*

Below: Stony Point 4-H Club, champions on exhibits at the Delaware County Free Fair in 1929. Pictured are teachers Mrs. May Benschoter and Mrs. Mamie Kindaide and the patrol coaches. *Photo courtesy of the Delaware County Extension office.*

Throughout the history of Delaware County 4-H, there has been a continued concern about the boys, girls, as well as patron coaches, known today as leaders or volunteers.

The earliest written documentation of the Delaware County 4-H Program was found in a dusty antique wooden file cabinet at the Delaware County Fairgrounds. The 1929 Narrative Annual Report of Extension Work, with its crispy, yellow-tinged and faded typewritten pages, was hand-bound in a brown folder containing a compilation of intriguing accounts of activities, philosophies, and documentation of days gone by.

Delaware County

Left: The 4-H members attend an achievement day program and judging school. Attending were 126 members and fourteen adults. *Photo courtesy of the Delaware County Extension office.*

Below: The largest Extension meeting held in Delaware County had over 200 people in this group representing six different school districts. *Photo courtesy of the Delaware County Extension office.*

E. A. Kissick, a county agent, reported on March 13, 1929, of twenty-eight clubs with 574 members: "4-H Club Girls should complete their FOOD PREPARATION WORK BEFORE SCHOOL IS OUT… Crop Club Boys should secure their seeds, (PURE SEED) at an early date… REMEMBER THE BOY WITHOUT PURE SEED IS WORKING UNDER A BAD HANDICAP. WHERE THERE IS A WILL THERE IS A WAY."

In 1933, 4-H clubs focused on canning, sewing skills, and poultry judging. Nettii Sitz, Extension agent, reported: "Butler Club is made up mostly of Indians. The Indian agent has brought in some good livestock for the boys. This last week has been bad for the gardens. Vegetables are getting scarce and if we do not have some rain soon there will not be anything out of the gardens to can."

In 1941, clubs participated in the cake and egg show, mattress-making day, and making homemade soap. "The total enrollment in 4-H clubs for the year 1941 is 688, 348 of which are girls. Club enrollment varies in size… two clubs, Zena and Lowery, have 100 percent enrollment—every child in the school, who is old enough, is a member of the club," reported Extension Agent Mariee Callaway.

Barbara Denney, current Extension educator, reported: "In years past, most of the 4-H clubs were hosted in schools. In 2009, the Delaware County 4-H consists of sixteen clubs with 350 members; however, only three are in-school clubs."

Dewey County

Contributed by Jean Bailey and Letha Crispin

There were twenty-six members in the 1954 Dewey County delegation to the Oklahoma State 4-H Roundup. *Photo courtesy of the Dewey County Extension office.*

The 4-H program began in Dewey County in the early 1920s under the leadership of James E. White, who was a county agent from 1924 to 1930. The first home demonstration agent, May Traver Wren, joined the county in 1926. In the 1930s and 1940s, poultry and home improvement were popular projects. By the late 1970s, there were 265 members in eight clubs with thirty-seven major projects to choose from.

Rene Crispin, a former member, submitted an article for the Dewey County Historical Society to print in its book *Spanning the River*, volume II. The article tells of the first Dewey County delegation to 4-H National Congress in Chicago. The 1925 delegation included Pearl Fairchild Dunlap, Della Livingston, and Willa Garlock.

Pearl tells this story: "A group of Oklahoma delegates were touring the stockyards when a cow got loose. I was a good target since I was wearing a red coat, but I wasn't in any danger since the Oklahoma boys made short work out of the cow by tackling her. (This exciting event made all the Chicago newspapers at the time.)"

Through the years, Dewey County has had numerous members receive first-place awards at the state and national levels. For the past fifty-eight years, seventy-three Dewey County members were inducted into the Oklahoma State 4-H Key Club. In 1967, the Dewey County 4-H Hall of Fame began to honor outstanding senior members. There are now seventy-seven members who have their pictures displayed on the wall at the fairgrounds.

Currently, there are 200 members in seven traditional, four Cloverbud, and two countywide project clubs. The 4-H educator responsibilities are shared by Mike Weber, agriculture educator and CED, and Jean Bailey, family and consumer sciences educator. In addition to the traditional projects, shooting sports and performing arts have become popular.

Dewey County

Dewey County members placed second in the county or school group of five steers at the 1959 Tulsa State Fair. They are (L-R): Mickey Vanderwork, Rada Sue Williams, Paula Craig, James Cary, and Joe Craig, showing Mickey Vanderwork's steer. *Photo courtesy of the Dewey County Extension office.*

There were eighteen members in the 2006 Dewey County delegation to the Oklahoma State 4-H Roundup, including one state officer and three project book medal winners. *Photo by Todd Johnson, Agricultural Communications Services.*

Ellis County

The Roundup winners of the Talks and Demonstration Contest in 1985 were Dana Roper, Brenda Harriman, Susan Nine, and Cindy Nine. *Photo courtesy of the Ellis County Extension office.*

Ellis County is a rural county with an agriculture-based economy. Currently, there are seven traditional clubs in the Arnett, Fargo, Gage, and Shattuck communities. Records indicate that John Bunyard was the first county agent, serving from 1916 to 1918. Mamie Chamberlain served as Ellis County Extension secretary for fifty-two years, from 1949 to 2001. She was recognized as the state's longest-serving employee.

Key Club awards have been presented to fifty-five members since it was organized in 1950. The first two Ellis County Key Club members were Beverly Edwards and Jay Wieden in 1950. Ellis County 4-H produced three national winners in 1941, 1946, and 1979. Also, the first female state 4-H president in Oklahoma was from Ellis County, Marla Johnson, from 1981 to 1982.

The Ellis County 4-H organized county Wonders of Washington (W.O.W.) trips to Washington, D.C. The first Wonders of Washington trip was taken in 1991. Since then, three more county trips have been enjoyed by over 216 family members and 4-H members in 1994, 1998, and 2003.

Travis Bowman served as a volunteer leader in Arnett and Ellis County for over thirty-five years and was awarded the State Honorary Membership Award in 1999 at State 4-H Roundup in Stillwater. Ellis County junior horse judging teams won state awards from 1994 to 1999, including in contests in Kansas and Texas. The 4-H youth participated in three bi-county textile tours to Texas and have had several awards in Make it With Wool at the state and national levels. The Fashion and Fabrics Project Club refurbished the jury room for the county's courthouse.

The 4-H youth continue to demonstrate a strong foundation in traditional 4-H projects like livestock, shooting sports, food and nutrition, citizenship, and fabrics and fashion. Numerous awards are won each year at the district, state, and even national levels in these project areas.

Ellis County

Above: The State 4-H Roundup delegation from 1961. *Photo courtesy of the Ellis County Extension office.*

Left: Nancy Davison and Donna Jo Dickson presented the Key Club Awards in 1962. *Photo courtesy of the Ellis County Extension office.*

Garfield County

Paul Hoover (1909–2009) was one of the first 4-H'ers in Garfield County and was a longtime 4-H supporter. The Hoover Building located on the Garfield County Fairgrounds was named in honor of Mr. Paul Hoover. *Photo courtesy of Louise Bradley*.

The 1942 Garfield County 4-H Roundup delegation with the girls wearing the official 4-H dress. *Photo courtesy of the Garfield County OSU Extension office*.

Garfield County Boys' and Girls' Agricultural Clubs were established in 1916 with clubs emerging throughout all the rural country schools. A demonstration club organization was in place three years before 4-H clubs became official. The first 4-H leaders were Mrs. Frank Seapy and Fred Moehle. Boys' lessons concentrated on poultry, farm animals, and gardening. Esther Herbert, who was a 4-H member in 1926 and 1927 at Reynolds School, stated that girls did a lot of gardening, cooking, and embroidery work. Among the first 4-H members were Joe Ben Taggart, Carolyn Rathmel, John Luckert, and Paul Hoover. Paul Hoover's first projects were geese and a Shorthorn steer named Victor Snowball. They had no trucks to haul cattle, so Paul walked his steer two miles into town and shipped it to Oklahoma City to the Southwestern American Livestock Show via a Frisco railroad car.

Garfield County

The 2002 4-H Roundup delegation. There was no longer an official 4-H dress and the delegates seem to be younger. Chaperoning were longtime educator and County Director Ron Robinson, 4-H Extension Educator Cindy Conner, and Program Assistant Mary Kate Reading. *Photo courtesy of the Garfield County OSU Extension office.*

Garfield County 4-H'ers have produced many state and national champions with their beef, sheep, poultry, and swine projects. Most recently added to the livestock show is the meat goat project. We have had a winning tradition for many years in meat judging, producing several state and national winners.

With the assistance of outstanding volunteers and agents, Garfield County 4-H has produced thousands of young people who have gone on to become community leaders and goal-oriented citizens. The rural towns of Waukomis, Bison, Garber, Covington, Douglas, Kremlin, Carrier, Hillsdale, Lahoma, Hunter, Pioneer, and Enid have passed 4-H family traditions down to many of our current members by telling stories of their great-grandparents' 4-H experiences. We have had the honor of hundreds of state record book project winners as well as numerous national winners. Seven state 4-H presidents have come from Garfield County: Burl Winchester (1932–1933), Aaron Gritzmaker (1940–1941), Melvin Semrad (1955–1956), David Semrad (1960–1961), Bruce Robinett (1961–1962), Missy Conner (1997–1998), and Dusty Conner (2002–2003). Garfield County 4-H members continue to carry on the tradition of providing community services that truly "makes a difference" in our county.

Mattie Cozart and Dorothy Anderson are giving their blue-ribbon demonstration on "Care of Clothing on Laundry Day" in 1940. *Photo courtesy of the Garfield County OSU Extension office.*

Garvin County

Contributed by Sarah Jolly

The Garvin County Cooperative Extension Service has played an important role in the lives of families in the county since around 1914, when George Lee became the first county agricultural agent. Many other county agents followed Mr. Lee, but it was not until 1946 that an assistant was hired to help with the 4-H program.

The next year, a young girl named Linda Butler joined 4-H for the first time. Linda was a member of 4-H in Pauls Valley. Within the next few years, Linda's two younger sisters, Barbara and Neva Sue, also joined 4-H. The sisters' main projects were canning, clothing, swine, and cattle. The Butler girls were delegates to 4-H Roundup. Linda was a delegate to Kansas City and Barbara was a Garvin County Hall of Famer. Linda cites 4-H summer camp as the absolute highlight of every year.

The Butler sisters and the Angus steers they raised for the state fair. *Photo courtesy of the Garvin County Extension office.*

Linda was a ten-year 4-H member and went on to marry W.C. Pesterfield. When Linda and W.C.'s children turned nine, they each joined 4-H, just as their mother had. As the children's 4-H project, the Pesterfields planted and sold sweet corn. The children sold enough corn each summer to put money away in a college savings fund. By the time they graduated high school, they each had enough money from selling corn to completely pay for college. Jason Pesterfield, Linda's son, was a Garvin County Hall of Famer and went on to be a major force in the early 1990s in the University of Nebraska's football program. Jason credits his 4-H experience for his success.

Today, the Pesterfields' sweet corn is still growing strong. Everyone in the county and surrounding area looks forward to summer when fresh-picked corn is available at their stand. Linda says the grandkids are coming now in the summers to help work the 4-H project they started nearly thirty years ago.

Grady County

Grady County 4-H began in 1921 with seven 4-H "centers." By 1930, twenty-three 4-H clubs were organized. In 1931, 385 boys and 396 girls were enrolled in the county program. Club work consisted of a county rally, picnic, team demonstration contest, appropriate dress contest, health contest, and club camp. Major projects included gardening, canning, poultry, and clothing. Clubs were scattered throughout the county and were associated with rural school districts (Annual Narrative Report, 1930).

The 4-H program has continued to grow throughout the past eighty-five years. Many 4-H members show livestock and participate in traditional rural agricultural projects. Grady County members have held district and state 4-H offices, won state and national record book competitions, and have been selected for the Blue Award Group (top ten 4-H members in the state).

Two 4-H members from Grady County have won the distinguished honor of being selected for the Oklahoma 4-H Hall of Fame. John Hancock, a Grady County 4-H member, served as Oklahoma 4-H president from 1963 to 1964.

In 2009, 1,000 4-H members were enrolled in eleven clubs. Clubs are located in Minco, Tuttle, Bridge Creek, Amber-Pocasset, Chickasha, Verden, Ninnekah, Alex, and Rush Springs. In addition, Glory 4-H and Spurs and Saddles 4-H were organized to serve special interests.

Grady County 4-H alumni include educators, business owners, corporate leaders, and state legislators. Many present 4-H members are children, grandchildren, and great-grandchildren of county 4-H alumni.

Grant County

Contributed by Robyn Rapp, Beth Peters, and Scott Price

The 1948 Grant County 4-H yearbook listed fourteen local clubs. They were the Deer Creek Boosters, Busy Bees, Mayflower Champs, Gore Go Getters, Nash 4 Leaf Clovers, Grand Valley Peppers, Medford Cardinals, Loyal Workers, Wakita, Manchester Clover, Pond Creek Home Front, Lamont Busy Bees, Jefferson Willing Workers, and Busy Sunbeams.

In 2009, there were four community clubs, Deer Creek-Lamont, Pond Creek-Hunter, Medford, and Wakita, with an enrollment of 128 members served by eighteen certified volunteers. There was also one Cloverbud club with twenty-one members.

Above: The 1935 4-H Roundup delegates. Front, L-R: Rose Milligan, Nelda Shiva, Maxine Biby, Enid Lacy, Helen Whitney, Alta Boyd, Marjorie Jones, Mildred Mitchell, June Albright, Margie Dixon, Alice Bobbitt, and Allen Williams. Back, L-R: Jack Summers, Jack Crider, Fred Cunningham, Lyndel Trenton, Paul George, Parker Goldsmith, Norman Lacy, Dwite Serviss, Clifton Honeyman, and Eldon Albright.

Right: The words "4-H CAMP" were spelled out with campers on July 26, 1990, during Grant/Garfield County 4-H Camp at Wentz Camp at Ponca City. The picture was taken from the top of the tower in the middle of campground. *Photo courtesy of the Extension office staff.*

Grant County

One of the earliest 4-H pictures in the Grant County Extension office is of the group who went to Roundup in 1935 (at left). The home demonstration agent was Rose Milligan and also attending was Allen Williams, the assistant county agent.

Since 1950, eight members have received the honor of membership in the Oklahoma Key Club. Grant County 4-H'ers have been members of the OKAY Performing Troupe. One member, Natalie James, received the presidential tray award.

Several members have been elected to the Northwest District Officer team and to the State 4-H Officer team. Natalie James was state 4-H president in 1989.

Over the years, 4-H'ers have had the opportunity to travel to activities and competitions in Washington, D.C., Chicago, Kansas City, and Denver.

Popular projects in 2009 were photography, shooting sports, food, nutrition and health, meat judging, and global information systems.

Special interest clubs formed from project work include a shooting sports team, Junior Master Gardeners, the Fashion and Fabric Sewing Club, and the Grant County 4-H Photography Club.

Grant County 4-H members continue to participate in county and district events and contests like Share the Fun, speeches and illustrated presentations, clothing contests, plant identification, baking, livestock judging, land judging, meat judging, the National 4-H Week Photo Contest, trap shoots, county and state fair, Roundup, and tri-county 4-H camp.

Oklahoma Cooperative Extension Service staff for Grant County on April 29, 2009. L-R: Robyn Rapp, Extension educator, Family and Consumer Sciences/4-H Youth Development; Scott Price, Extension educator, Agriculture/4-H Youth Development and CED; and Beth Peters, senior Extension secretary. *Photo courtesy of the Extension office staff.*

Earl Duane Hawkins with his Grand Champion Angus steer at the Grant County Free Fair in 1956. *Photo courtesy of the Extension office staff.*

Greer County

Contributed by Trena Medlock, Carol Toole, Glynadee Edwards, and David Raby

The year 1909 appeared to be the beginning of a very productive time in Oklahoma. Greer County had a brand new courthouse, county commissioners announced the first ever Greer County Free Fair, and Oklahoma 4-H found its humble beginnings.

Mr. James David Raby, 4-H member and supporter since 1948. *Photo courtesy of the Greer County Extension office.*

Early on, 4-H clubs met in school. In 1910, Greer County had an astounding fifty-two school districts, while today there are two. Many families earned a living by means of agriculture. The 4-H program brought informative instruction to youth in Greer County on farm safety, food preservation, textiles, crop production, livestock production, and poultry production. During a 4-H livestock exhibition, one particular event was swine-breeders. Members would exhibit their top sows and boars. In the swine show, pigs were exhibited as singles or by truckload or boxcar.

During an interview with David Raby, a leader and longtime friend of 4-H, we learned that he was awarded, through 4-H, the prestigious Wheat King award in 1956, one of only two in the state. To win, Raby exhibited one peck of wheat. Each grain in the peck matched in size, was sifted by wind and screens, then examined under a magnifying glass to ensure that there wasn't a speck of chaff. Raby traveled to Chicago to receive his award. While there, he toured the Chicago Board of Trade where wheat is bought and sold.

When asked what impresses him most about 4-H, Raby replied: "4-H has held true to the basic ideals of HEAD, HEART, HANDS, and HEALTH; teaching youth to be responsible for themselves. Notably, past members return to the skills they learned in 4-H and are able to give back to others by helping them achieve their goals."

Oklahoma 4-H Congress in 1966, Chicago-bound. *Photo courtesy of the Old Greer County Museum.*

Harmon County

Originally written by Donnie Audrey Kite Cooper

In the early years of 4-H clubs in Harmon County, boys and girls joined at age ten and met once a month with the county agents present. The members were taught to use Robert's Rules of Order for conducting business meetings.

Members learned to sing familiar youth songs with one elected to lead the music, the Pledge of Allegiance to the U.S. flag and other group responses such as the meaning of each of the four H's, which are Head, Heart, Hands, and Health (formerly Home).

"Uncle" Tom Marks organized the Corn Club where he worked in Texas. He moved to Hollis to work with the 4-H clubs in Harmon County and strengthened the work for several years. After years of service, Marks was elevated to a position with the Extension Service at Oklahoma A&M and moved to Stillwater.

LaHoma 4-H Club members view some Leghorn chickens during a club tour. *Photo from the Extension annual report, 1939.*

County Agent Clarence Burch talks to club members at the annual county 4-H picnic. *Photo from the Extension annual report, 1937.*

The family and consumer science educator of the time challenged the girls to select projects to complete for home use. Likewise, the agriculture educator encouraged boys and girls to select projects they could begin and finish. Competition was encouraged. The demonstration was a method of presenting information. Almost any subject lent itself for demonstration. This could be a single member or two or more working together to present information.

Great incentive to achieve as an individual could be found through the awarding of trips to annual 4-H Roundup, the annual state fair in Oklahoma City, and trips to the National 4-H Congress.

Training in 4-H clubs helped rural youth to learn more about the world outside of their county. Lifetime friendships were formed and respect for the talents of others made permanent impressions. The 4-H program was and is an important program for youth.

Harper County

Submitted by Carol Laverty and Vernor Bockelman

The first Harper County Extension office was established in 1917. The first 4-H club was called the Corn Club and was organized by Laura McClain. McClain was known for her Buffalo Bunny Sausage, as Harper County had many rabbits. Mrs. Sweet helped McClain and was the agent from 1918 until 1924. Harry Wheat was the first farm agent in Harper County. In 1918, Mr. Porter, a farm agent, helped Mrs. Sweet organize the first 4-H club. Harper County had no agents from 1926 to 1934 because of the Depression.

In 1934, Bill Bland, who was a county agent, took 4-H'ers on many trips. Vernor Bockelman and Miriam Bockelman were active 4-H'ers and attended the first 4-H Roundup. Roundup delegates stayed in homes all over Stillwater and attended meetings in tents.

In 1938, 1,500 people attended the Harper County Achievement Banquet, where Vernor Bockelman was the emcee for the evening. He was also selected to emcee the Oklahoma State Livestock Show banquet in Oklahoma City with several thousand people attending, including the governor of Oklahoma.

During World War II, Harper County had fundraisers to help purchase Bombers for the United States Air Force.

Vernor Bockelman received a $150 scholarship on his 4-H record book that enabled him to attend Oklahoma State University.

Harper County's 425 4-H members and their parents organized a countywide club federation in 1948 and held its monthly meetings as a group so that 4-H'ers across the county could get to know each other better. Max Barth, a county agent, said the results from this activity would be reflected in county fair and junior livestock shows. He also said interest in record books would be stimulated. More than 200 people attended the meetings.

Haskell County

Contributed by Brian C. Pugh, Shawna Hudspeth, and Lisa McRay

The Haskell County Award Banquet scholarship winners in 1955. *Photo courtesy of the Haskell County Extension office.*

Although the exact history of Haskell County's 4-H program is unknown, there is an abundance of evidence that proves that 4-H has been a longstanding tradition. Archived in the county Extension office are old photographs and news articles, some dating back to the 1930s, that share a glance at the county's 4-H past. The thing that has remained unchanged is the involvement of Home and Community Education (HCE) groups, once known as home demonstration. Years ago, the girls in the 4-H clubs were very active in sewing. Photos

Haskell County

show fifth-year sewing projects of dresses and aprons. Today, we occasionally see a tote bag or a pillowcase, but sewing is approached as a craft instead of a necessary skill. Other photos show boys participating in judging contests and articles tell of them being heavily involved in conservation, agronomy, and livestock projects.

Livestock projects are still among the most traditional for a county based on farm living. Generations of 4-H families are still operating the farms that their great-grandparents began. The livestock program also gains more community financial support than any of the other 4-H programs. With the county's southern border touching the Ouachita Mountains, forestry has also been a popular 4-H project. During the 1990s, the county became active in the state shooting sports program. Through both of these project areas, many youth from the Haskell County Club have been able to compete in local, state, and national competitions, where they've been recognized as winners in national forestry and national shooting sports in recurve.

While Haskell County is obviously a very traditional 4-H program, there are many newer opportunities for youth as well. Technology has become a major part of our world and the county 4-H program is no exception. Today, youth are building rockets, learning digital photography, creating GPS maps, and keeping in constant contact with the people they meet in 4-H through social networking. Haskell County 4-H is proud to honor its past members, celebrate the present program, and envision the future of 4-H's involvement in our community.

The Haskell County forestry team attended national forestry competition. *Photo courtesy of the Haskell County Extension office.*

The 4-H members at rocketry day camp. *Photo courtesy of the Haskell County Extension office.*

Hughes County

Contributed by Ashlan Wilson

You can ask just about anyone from Hughes County if they were in 4-H as a young person and more than likely their answer would be yes. Many people have interesting stories and fond memories of 4-H and how the program influenced them in a positive way.

The 4-H program has been an important part of many lives in Hughes County. Since the very beginning of 4-H clubs in Hughes County, the program has helped young people "To Make the Best Better." Not only have members gained from their experiences in 4-H, but the program has involved the entire family.

Today in Hughes County, there are seven clubs that serve 258 members in five communities. There are twenty-three certified 4-H volunteers and many other adults who volunteer their time to the 4-H program. Hughes County 4-H'ers are involved in many projects including beef, swine, leadership,

Above: Beverly Chapman, Oklahoma 4-H Leader of the Year in 1998, and Laramy Wilson, the 2006–2007 state 4-H president. *Photo courtesy of the Hughes County Extension office.*

Left: The 2008–2009 4-H officers and Extension educators. Front row, L-R: Jeffery Hardwick, Megan Brigden, Ashlan Wilson, and Kevin Meeks. Back row, L-R: 4-H Extension Educator Aubie Keesee, Ross Turner, Christian Tollett, and 4-H Extension Educator Robyn Jones. *Photo courtesy of the Hughes County Extension office.*

Hughes County

County Agent Jess Barbre with Spaulding 4-H member Tommy Kendall. The photo was taken around the mid-1950s. *Photo courtesy of the Hughes County Extension office.*

shooting sports, foods, and many others. The 4-H'ers are active on the local, county, district, and state level with projects, trips, contests, and community service. There have been numerous national and state project winners from Hughes County. In addition to many district and state officers, Hughes County is home to two Oklahoma 4-H presidents: Henry Wolf (1944–1945) and Laramy Wilson (2006–2007).

The Hughes County 4-H Program has been directed by many great Extension educators throughout the years. This dedicated group includes the late Jess Barbre and Allen Burns, as well as Monroe Sumter, and the current Extension educators, Aubie Keesee and Robyn Jones.

Our certified volunteers have dedicated years of service to the 4-H program. Beverly Chapman was named Oklahoma 4-H Leader of the Year in 1998 and Debbie Wilson received the 4-H Lifetime Volunteer Award in 2005.

Hughes County 4-H is happy to celebrate the one-hundredth birthday of 4-H and is looking forward to the next one hundred years.

Jackson County

Contributed by Mitzi Pate, Extension educator, and Nolynn Moreau, Jackson County 4-H reporter/ambassador. Information obtained from Edna Louise Richeson Hall, Larry Derryberry, Robetha Ann Masters Darby, Martha Jane Howard Sauer, Robert Carroll Welch, Frederick Roe Worbes, Roy Wayne Moreau, and Brent Scott Howard

The earliest record of a 4-H meeting in Jackson County is located in the Prairie Hill School's 1941 yearbook, *The Ranger*: "On January 8, 1941 the 4-H club of Prairie Hill was organized with 45 members enrolled." The article states that various project areas—food preparation and preservation, clothing, home improvement, yard improvement and horticulture, farm accounts, poultry, dairy, handicraft, livestock, and health—had elected captains. Elected officers were President Norman Kaufman, Vice President Joyce Chenault, Secretary Wanda Wilcox, and song leader Junior Riddle. Club leaders were referred to as coaches.

It is believed that 4-H was organized in Jackson County prior to 1940, as the first county agent, Thomas A. Sheriff, was appointed on February 12, 1912.

Projects in every area from animal science to science and technology have been completed by Jackson County 4-H members. The most popular 4-H project area has always been public speaking. Jackson County 4-H members from 1945 to 2000 have claimed "public speaking and social skills" as the most valuable skills they learned in 4-H. Life skills such as sewing, cooking, responsibility, record keeping, and teamwork are also valuable skills learned.

Popular 4-H events in Jackson County are livestock shows, speech contests, dress revue, summer camp, county fair, achievement banquet, the poster contest, and Share the Fun. Events from the past that were popular were judging schools, food contests, demonstration contests, Robert's Rules of Order, and county 4-H exchange programs.

District, state, and national trips such as Youth Action Conference, State 4-H Roundup, National 4-H Congress, and Citizenship Washington Focus are often the "most memorable events" for county 4-H'ers. Winning project awards, being elected county officer, and being selected for the Jackson County 4-H Hall of Fame or state 4-H Key Club are favored "highlights" of any member's 4-H career.

Jefferson County

Jefferson County has many exceptional 4-H alumni. One outstanding individual is Phillip Scott, a leading citizen in Jefferson County. Scott credits 4-H with the decision-making abilities he uses every day as a successful cattle rancher and prominent attorney in southern Oklahoma. He attributes being able to effectively make choices to his national 4-H livestock judging experiences.

As a ten-year member of the Waurika 4-H Club, Scott served in many capacities, including president of his local club as well as Southwest District 4-H vice president, and received recognition for his many 4-H activities as a runner-up for the Oklahoma 4-H Hall of Fame in 1961.

"Juggling 4-H, athletics, school, and farm duties was just part of what we did as kids," says Scott. "The fall of 1960, I was a member of two successful teams—a nationally competitive livestock judging team and a high school football team competing for the state championship, events all occurring in the same week and even on consecutive days."

Scott and his teammates Jack Savage, Bill McGowan, and J. R. Dunn competed in the 1960 Oklahoma 4-H livestock judging contest. Thanks to the efforts of many people including his coach Jim Timmis, assistant county agent and judging coach, his team won the contest. It entitled them to compete in the Kansas City Royale competition. Savage took the National Beef Judge title and Scott received the top honor in national sheep judging.

Earning many honors, Scott says that a 4-H scholarship enabled him to attend Oklahoma State University. After graduation, the accomplished student acquired a law degree from the University of Oklahoma. Among his many accomplishments, Scott was awarded the General Hal Muldrow Pistol Award for his outstanding military leadership. Over his military career, he graduated from Ft. Holabird's Army Military Intelligence School and was awarded two Bronze Stars for his courageous bravery in Vietnam.

Johnston County

In 1909, Oklahoma 4-H was born in Johnston County in the town of Tishomingo. The first corn club was organized with fifty Johnston County boys by W. D. Bentley to promote the growth and improvement of corn. The February 8, 1910 edition of the *Johnston County Capitol Democrat* announced fair prizes of five dollars for the best white and best yellow corn exhibited and a pony would be given to the boy who raised the best ten ears of corn.

W. E. Brogden was one of the original leaders of the corn clubs. His daughter, Vera Taylor, followed his example by becoming a home demonstration agent, serving in Johnston County for nineteen years. Their legacy is a strong tradition of 4-H club work that is still in effect today with twelve clubs and 604 members.

Irene Combes Cooper, a 4-H member in the forties who later served as an assistant home demonstration agent, shared this information: "From the early years until the early seventies, boys had agriculture projects and girls did cooking and sewing." Even though it was unusual for girls to have agriculture projects, Mrs. Cooper raised a hog and since there was no fair barn, showed it in a panel ring on the Main Street of Tishomingo. Jane Ferris, daughter-in-law of charter Corn Club member Ples Ferris, recalls that the men of the community donated calves and pigs to be auctioned to raise funds for the first

Lela Ferris, wife of charter Corn Club member Ples Ferris, taught sewing at Connerville School every Thursday afternoon for ten years. Fashion revue participants from her 1961 club included Elaine Ferris, Lillian Kellage, Sharon Wallace, Marva Mullins, Mary Holt, Margie Coles, Sue Brown, and Linda Coles. *Contributed by Jane Ferris*.

The award winners from the 1966 4-H Achievement Banquet. The green 4-H jackets were awarded by the Home Demonstration Club for grand champion speeches, a tradition still in effect with current HCE clubs, leading to a very strong speech program in the county. *Contributed by Vera Taylor*.

Lela Ferris pictured with Home Demonstration Agent Vera Taylor.
Contributed by Jane Ferris.

fair barn. The county has recently completed a 30,000-square-foot fair barn facility, which also provides a permanent home for the OSU Extension office and 4-H program.

Popular current projects are recreation, shooting sports, and photography supplementing the traditional agriculture and home economics projects. Leadership development is a major impact of 4-H involvement. The program has been represented by state song leaders Melvin Mitchell from Mill Creek and Jack Baker from Tishomingo, state reporter Star Smith Edwards from Wapanucka, and our current state 4-H representative Shane Jemison. Ambassadors from Johnston County include Star Smith Edwards and Tori McCollom from Tishomingo and Kyle Foster from Wapanucka. Our first state ambassador was Dana Smith from Coleman. Dana was killed in an automobile accident in 1996 and a state scholarship has been established in her name.

Wayne Easterwood is pictured at a livestock show in 1948.
Contributed by Wayne Easterwood.

Kay County

Contributed by Rachael Smith and Larry Klumpp

Roundup delegates from 1952. *Photo courtesy of Kay County Extension.*

"I pledge my head to clearer thinking, my heart to greater loyalty, my hands to larger service, and my health to better living for my club, my community, my country and my world."

If you are currently an active 4-H'er or 4-H alumni, the words to the 4-H pledge are forever embedded into your heart and mind. Past to present, the 4-H program across the nation and especially here in Kay County has helped put values and morals into the lives of children.

Ray Shiltz, a 4-H alumnus, stated that 4-H was a major part of his life growing up in Kay County. Back in the 1940s, the main projects were agriculture and home economics-based. Shiltz reminisced about "the good ole times" he and fellow Kay County 4-H'ers had showing their livestock and participating on a judging team. He spoke about the time he won the impressive dress contest on two different occasions and how he traveled to

Kay County

Roundup delegates from 1951.

Kansas City and Chicago to judge livestock. Back then, older 4-H'ers looked forward to the State 4-H Roundup and being able to participate in various speech contests. Shiltz also commented on how the towns in Kay County were very involved in supporting the 4-H organization.

Past to present, many things which make 4-H such a wonderful organization have not changed. Today, 4-H projects are still based on agriculture and home economics, but they now have arts and nature projects as well. The 4-H program has an area for every youth in Kay County to become involved.

The 4-H members prepare for the 1953 Oklahoma City Junior Livestock Show.
Photo courtesy of Kay County Extension.

Kay County

The first-place team in cattle grading at the junior livestock show in 1953. *Photo courtesy of Kay County Extension.*

Oklahoma delegates for Club Congress in 1952. *Photo courtesy of Kay County Extension.*

Kingfisher County

Contributed by Ernest Hellwege, Keith Boevers, and Valeri Evans

Kingfisher County members participated in annual exchange trips to learn about the culture and 4-H projects in other states. This idea grew to become an international opportunity. Ernest Hellwege, a Big 4 4-H member and state 4-H president, was selected as one of seventeen 4-H'ers from across the nation to participate in the first International Farm Youth Exchange Program (IFYE) with seven countries of Western Europe in 1948. This opportunity is still available and is now called International 4-H Youth Exchange.

A highlight for every 4-H member is to attend State 4-H Roundup on the OSU campus. Nine 4-H'ers from Kingfisher County attended in 1947 along with their assistant county agents. *Photo courtesy of Ernest Hellwege.*

Kingfisher County 4-H records date back to the 1930s with fourteen clubs across the county: Alpha, Altona, Big 4, Cashion, Dover, Greenwood, Hennessey, Lacey, Lone Star, Loyal, Omega, Red Valley, Square, and Willing Workers. County officers met regularly. This group was known as the County 4-H Club Federation.

Today, Kingfisher County has six clubs representing Cashion, Dover, Hennessey, Kingfisher, Lomega, and Okarche. There are 491 members in every project area, with many other youth involved through non-traditional means. Members participate in fashion revue and impressive dress, camp, speech and illustrated presentation contests, junior Roundup, achievement banquet, Share the Fun, and many other educational events. The horse club and shooting sports groups are organized project clubs and are very active on the county, district, and state levels. Kingfisher County is well represented on the state level with project winners, state ambassadors, and members of the state leadership team.

The 4-H events included the county rally, camp, dress revue, demonstration contest, the county banquet, and others. At officer elections, county project captains were appointed to lead project areas including food preservation, food preparation, clothing, home improvement, livestock, dairy, poultry, crops, and horticulture. Kingfisher County boasted the state winning livestock judging and poultry judging teams for several years, plus many state project winners.

Kingfisher County

Kingfisher County was known for being the "team to beat" in state competition. In 1946, Assistant County Agent L. J. Cunningham coached the state winning poultry judging team, which consisted of Ernest Hellwege, Donald Hellwege, and Tom Birkes. This team went on to compete on the national level. *Photo courtesy of Ernest Hellwege.*

Kingfisher County's 1953 Share the Fun performance at Roundup was chosen to be considered by the National Committee, which selects the talent to perform at the Share the Fun breakfast. This photograph was taken at the 1953 Roundup in the building now known as Gallagher-Iba Arena. *Photo courtesy of the Kingfisher County Extension office.*

Kingfisher County

Kingfisher County compiled the photographs in this collage to show the variety of 4-H opportunities offered to youth in the areas of agriculture and home economics. This collection of photographs was taken circa 1945–1965. *Photo courtesy of the Kingfisher County Extension office.*

Kiowa County

Contributed by Frances Davis Coffey, Allen Moore, Brenda Medlock, and Kent Orrell

Around 1909, many boys began joining Kiowa County Corn Club. Led by an unknown man, young men learned how to plant, grow, harvest, store, and utilize their crops. Soon after, girls started gathering together, exchanging canning techniques and learning how to garden, sew, preserve food, and take care of their households. By 1914, 4-H programs had emerged and the boys and girls were meeting together in their own community/school clubs. There were 112 schools in Kiowa County in 1914 and each had its own 4-H club. The 4-H program had arrived and allowed boys and girls opportunities to work, study, learn, and become leaders in their communities, counties, and state. In 1946, Kiowa County 4-H was led by a young man named Ted Davis. As county president, he set goals for himself and soon reached one of those goals. Davis was elected state 4-H president in 1948 and set the standard for other young people in Kiowa County.

Those standards still exist today. Although most of the forty-eight communities have dwindled, young people in Kiowa County have

Kiowa County delegates to Roundup in 2008. Back row, L-R: Scott Harris, Katie Blevins, and Ben Smith. Next row: Adam McKay, Ashley Bredy, and Shelby Cook. Front row: Korinne Medlock, Casey McKay, Krischa Medlock, Brittney Burton, and Hannah McCollom. *Photo courtesy of the Kiowa County Extension office.*

always had the opportunity to experience growth through their local, county, and state 4-H programs. Today, there are four community clubs in Kiowa County. These clubs include Snyder, Mountain View, Hobart, and Lone Wolf. Each allows 4-H members to meet and exchange ideas. Today, the main topic at a meeting may be how corn has been genetically modified rather than how to produce and store it. Youth in Kiowa County still have the opportunity to grow and develop their skills through 4-H.

Frances Davis demonstrates her electric project. Her exhibit was State Individual Electric Demonstration winner in 1949.
Photo courtesy of the Kiowa County Extension office.

Kiowa County

Kiowa County delegates to Roundup in the spring of 1950. *Photo courtesy of the Kiowa County Extension office.*

Ted Davis, from the Cooperton 4-H Club, was state 4-H president in 1948. *Photo courtesy of the Kiowa County Extension office.*

Latimer County

Left: Former Latimer County Extension Agent Joe Jeffrey presents a young member with an award packet. *Photo courtesy of the Latimer County Extension office.*

Below: Latimer County 4-H member Heather Edington performs at the 4-H Veterans Day honor program. *Photo courtesy of the Latimer County Extension office.*

Latimer County is nestled on the south side of the San Bois Mountains. It sits between Pittsburg County on the west and LeFlore County on the east in the southeastern part of the state. With only two incorporated towns, Wilburton (the county seat) and Red Oak, and four schools located within its 729.12 square miles, Latimer County has a population of just over 10,000 residents.

After it was discovered in the 1870s that Latimer County was rich with coal, the county's early economy was mainly based upon this mineral. Coal mining suffered a decline and a collapse in the 1920s due to labor disputes and the rise of petroleum as a fuel as well as the onset of the Great Depression. Today, Latimer County is still known for its rich fuel resources, only now it is known for its natural gas deposits.

To non-residents, Latimer County is famous for its "Robber's Cave State Park," located just five miles north of Wilburton. Mention to someone that you are from Latimer County and they

usually don't recognize where that is, but mention Robber's Cave State Park and they usually reply that they have been there! Robber's Cave was a popular hideout for Civil War deserters and outlaws such as The Youngers, Jesse and Frank James, the Dalton Gang, and Belle Starr. Visitors to the park can climb their way to Robber's Cave and see its caves and the natural rock corral that held the fugitives' horses from the sight of lawmen scouring the area.

Latimer County also has a rich history of hardworking individuals that have kept a struggling county alive. Latimer County 4-H has long had a part in helping to make those individuals strong for their adult years. Latimer County 4-H has a rich and thriving past in Latimer County and a future that looks even brighter!

Members of the "Bull and Crockett" 4-H Pork Team prepare to serve food during the annual Latimer County 4-H Pork BBQ Roundup. *Photo courtesy of the Latimer County Extension office.*

LeFlore County
Contributed by Amanda Beshear

Oklahoma 4-H, celebrating one hundred years! As LeFlore County celebrates the past one hundred years, we also look toward the future. LeFlore County has a rich, traditional 4-H history.

Regina Rogers-Wylie, a 4-H member from 1979 to 1986, participated in projects like citizenship and community service. She was a 4-H Key Club member and attended National 4-H Congress, Washington Focus, LeFlore County 4-H North to Alaska, and a ten-state citizenship/leadership tour. She loved having friends from all over the United States. Wylie is a family and consumer sciences teacher because of her 4-H experiences. Wylie's 4-H Agent Terri Stockstill and Agriculture Agent Jim Kelley were huge influences and were always there and willing to go the extra mile to help her succeed.

Other LeFlore County 4-H members recall 4-H being a huge part of the school. The 4-H meetings were held in the gym and every student in third through eighth grade was involved. Share the Fun was a huge county event. Clubs were open to kids from any area in the county. Members joined clubs that met their interests.

Currently, LeFlore County 4-H member Seth Hall is working to promote Dexter cattle. Twenty years ago, Dexter cattle were listed as critical by the American Livestock Breeds Conservancy, but today they are a recovering breed. Dexters were first introduced to LeFlore County by Seth in February 2009. Seth continues to exhibit his Dexters and educate people about the breed. He was the first person from LeFlore County to win Grand Champion Dexter Female and Reserve Champion Dexter Steer at the 2009 Houston Livestock Show.

LeFlore County 4-H continues to change and move in new directions. LeFlore County currently has ten clubs with forty-two volunteers serving approximately 725 youth. The future is bright for LeFlore County 4-H.

Lincoln County
Contributed by Ross Sestak

The 4-H Roundup delegates for 1937. *Taken from the accounts of Rodger Goodbary and Edna Ruth Goodbary's record as a 4-H club member.*

The Lincoln County 4-H Program can be traced back to the early 1900s. The Boys' and Girls' Club Federation of Lincoln County had the first opportunity to display their work during the Lincoln County Free Fair in September 1921. To participate in the fair, exhibits must have first been part of the district or community fairs. J. W. Guin was the county agent at that time. Many of the records from early 4-H work were lost in the Lincoln County Court House fire occurring on December 23, 1967.

Lincoln County

Several archived records have been preserved from the mid-1930s. In 1939, Lincoln County 4-H clubs increased from fourteen to twenty-one, representing a 50-percent increase from the previous year. There were 664 young people enrolled for that year. The traditional club work we know today did exist in that era. Many of the male 4-H members were actively engaged in soil and crop improvement. Terracing clinics were conducted in several portions of the county. The hope of saving and rebuilding the soil was given to boys for the continual dream of farming. These boys were responsible for terracing some 1,493 acres. The first farm to be terraced in the county was owned by Russ and Mattie Goodbary. One of their sons, William Allan, went on to be a decorated member of the 4-H program and served as state 4-H president in 1935. The traditional programs known to 4-H were instilled in the pioneers of Lincoln County and are still present today.

This photograph of a Lincoln County 4-H officer meeting is believed to have been taken between 1950 and 1951. *Courtesy of the Lincoln County Extension office.*

Built on a strong tradition that still exists today, members worked then and now "To Make the Best Better." Historically, Lincoln County is richly blessed with many district and state officers, state project winners, scholarship winners, and 4-H Hall of Fame recipients. In 2009, Lincoln County 4-H was 533 members strong, had a presence in eight Lincoln County communities, and was guided by thirty-three certified 4-H volunteers and the staff of the Lincoln County Cooperative Extension office.

The 4-H members prepare to load the bus to spend the week at 4-H camp. This group participated in 1958–1959. *Courtesy of the Lincoln County Extension office.*

State 4-H Roundup, 2008. Delegates to State 4-H Roundup are pictured with the Spirit Rider at the northeast corner of Gallagher-Iba Arena. *Courtesy of the Lincoln County Extension office.*

Logan County

Contributed by Cathy James, Gaye Pfeiffer, and Kathie Dellenbaugh

The first 4-H club in Oklahoma was in 1907 in Orlando, Oklahoma, in northern Logan County. It was started as a program for rural youth with the objective to create more interest in agriculture.

Competition was introduced in 1910 when the boys of Logan County grew an acre of corn to win an all-expense-paid trip to Washington, D.C. County rallies began by 1920, and in 1926, the county rally was held in Mineral Wells Park with 500 youth and 200 adults attending and included a dress revue for the girls. By 1927, there were nineteen 4-H clubs in the towns and rural areas of Logan County and the County 4-H Officers Federation was organized. In the forties, county rallies were held at Fogarty School, where 4-H'ers participated in demonstrations on timely topics.

The first county Share the Fun contest was held in 1951 with a category for individuals and groups in musical, dramatic, or novelty acts. Logan County 4-H Club members held their first camp in 1954 at Camp Redlands near Stillwater. The late 1960s brought the first county Make It With Wool, the Bake Show, the Flower and Vegetable Show, and the Safety Poster Contest. During the 1970s, Logan County 4-H clubs had numerous state, regional, and national winners, with 4-H members attending the National Congress in Chicago. Participation in Roundup was high in the 1980s and resulted in the introduction of a point sheet to determine attendees, which is still being used today.

Division winners in the Logan County Make It With Wool contest. L-R: Rene Bulling, Mulhall-Orlando, pre-teen; Terri Bulling, Mulhall-Orlando, junior dress; JoVanna Pfeiffer, Mulhall-Orlando, senior suit; Sharon Downey, Coyle, senior coat; Cathy Bennett, Crescent, senior dress; and Mary Furlong, Guthrie, junior suit. Sponsored by the Logan County Sheep Producers. *Photo courtesy of the* Guthrie Daily Leader.

The nineties marked a change in program emphasis from traditional agriculture projects to more urban programs and saw

Logan County

The 1971 Logan County 4-H livestock judging team. L-R: John Pfeiffer, Debbie Rigdon Hamilton, Billy Joe Lucas, and Cathy Bennett. *Photo courtesy of the Logan County Extension office.*

Logan County 4-H members attending the 2004 Northwest District Leadership Conference in Woodward. L-R, front row: Heather Young, Brandi Mack, and Andy Pfeiffer. Back row: Extension Educator Dave Williams, 4-H Leader Arleen Mack, Derek Stewart, and Megan Hickman. *Photo courtesy of the Logan County Extension office.*

a return of second-generation 4-H families to the program. The millennium brought a renewed spark to 4-H enrollment and a new emphasis on recruiting and training adult volunteers.

Traditional projects continue to be strong with a broadened focus on science and technology. Today's 4-H program focuses on challenging youth to actively participate in the learning process. Over 300 4-H members in Logan County belong to six organized clubs. These busy, productive members annually exhibit approximately 300 county fair exhibits, 120 state fair exhibits, and over 400 livestock exhibits. They actively engage in community service, speech contests, and other events that enhance their personal growth and leadership skills.

Love County

Contributed by Charles J. Sykora

The 1957 Love County 4-H banquet. *Photo courtesy of Chick Alexander.*

Love County has been blessed by the 4-H program which officially started sometime after statehood in 1907. There are Love County Free Fair ribbons dated 1925 awarded to at least one family. Early youth leaders and especially the Extension services personnel should be commended for the skills learned and overall development of youth through the 4-H program.

The present fairgrounds building in 1969 included facilities for showing animals as well as a building for indoor activities. Livestock shows in the early fifties were on Marietta Main Street, at the rodeo grounds, in Embry cattle pens on the east side of town, and on the auction barn parking lot on South Highway 77. The logistics of always having enough material and manpower to temporarily construct and dismantle enough pens showed the enthusiasm and importance of the event. Today, many indoor projects are held at different schools in the county as well as the county fair facility. In the past, the courthouse as well as churches and the old Marietta gymnasium also provided space for these events. These particular churches and the old gym structures no longer exist.

Projects and events as well have changed over the years to fit the needs of the times.

The 4-H youth from Love County at Roundup. *Photo courtesy of the Love County Extension service.*

Major County

Major County delegates for Roundup in 1949. *Photo courtesy of the Major County Extension office.*

The first record of the Major County Fair is 1930. During the 1930s, there were more than eighty schools in Major County. Families resided on eighty to 160 acres across the county. During the 1940s, many of the small schools consolidated into larger country schools. The country schools all had a 4-H club and were very active in the county 4-H contests. The county Extension agents had a very active role in teaching sewing, cooking, agronomic practices, livestock care, and judging. By the late 1960s, most of the remaining country schools had closed their doors and merged into the four main town schools.

Shown in the above picture is the Major County delegation to State 4-H Roundup in 1949. The 4-H programs really filled a need in the rural Oklahoma lifestyle. The communities had small local general stores. Families grew their own produce, raised the meat products, collected farm-raised eggs and, of course, had their own milk cows. The 4-H members and their families did not travel into town as much as families do today. The 4-H program provided families an opportunity to socialize and learn new skills. The 4-H meetings taught many of the same skills as they do today, such as public speaking, parliamentary procedure, judging skills, sewing, cooking, food preservation, gardening, and horticulture.

Currently, the four school districts in Major County are Fairview, Ringwood, Aline-Cleo, and Cimarron. Aline-Cleo is comprised of the Cleo Springs Community and the Aline Community. The Cimarron School is comprised of the Ames and Lahoma communities. There are currently 175 4-H members in Major County 4-H. Members still learn sewing, cooking, livestock care, judging, gardening, and horticulture. However, 4-H members have found new projects like shooting sports to be fun and entertaining. The 4-H members have a tough time scheduling around various school and church activities, as families are much more mobile than they were in the 1940s. In Major County, 4-H still teaches the value of helping others and the importance of taking responsibility for your actions, and provides youth with leadership opportunities and many opportunities to grow as they complete their 4-H projects.

The 1950 Major County dress revue winners. *Photo courtesy of the Major County Extension office.*

Livestock production and exhibition is a popular project area for Major County. *Photo courtesy of the Major County Extension office.*

Marshall County

Contributed by David Sorrel and Sara West

Marshall County's first annual camp, held in 1947 south of Shay, was noted as the "most successful 4-H activity" by County Agent Dale Ozment. Campgrounds host Reuel Little gave campers a cruise around Lake Texoma in his boat. *Photo courtesy of the Marshall County Extension office.*

The 4-H program has been embedded in the culture of Marshall County since 1909 and continues to impact the lives of youth and their families.

In the early days, Marshall County did not have a full-time 4-H agent, so the home demonstration agent and agriculture agent would conduct many of their educational programs on the farm and in the home. The 4-H clubs were quickly organized in schools around the county with clubs at Antioch, Archerd, Aylesworth, Buckhoit, Bowlin, Cumberland, Kingston, Lark, Lebanon, Lone Elm, McMillan, Shay, Simpson, Timber Hill, Tyler, and Willis.

In the beginning and for the better part of the twentieth century, the girls and boys met separately and participated in completely different project areas. Culturally, there was a distinct difference in the type of training for boys and girls at that time. Young women were taught food preservation techniques and young men learned better farming techniques. Most of the projects were to help families get above a level of simple subsistence. The 4-H members used cotton from a local cotton gin and made hundreds of mattresses for people who did not have one. Records at the Marshall County Genealogical Society talk about the Freedom Gardens that 4-H members planted during World War II to help ensure an abundant food supply.

Cattle and forage have always been, and continue to be, the mainstay of Marshall County's agriculture economy. In 1947, soon after Lake Texoma was built, Marshall County 4-H members organized their first overnight camping experience. The 4-H Camp at Texoma Christian Camp became an annual tradition that lasted sixty-one years. In 2009, for the first time, 4-H members attended camp at another facility, Camp Classen in the Arbuckle Mountains.

Marshall County

Today, Marshall County reaches over 700 youth through school enrichment and traditional 4-H club activities. Marshall County has six organizational clubs that meet after school at Kingston and Madill, with animal science, dog club, DIY, food science, horse club, and shooting sports meeting on a monthly basis. Youth are actively involved in club activities and many traditional county events such as talks and demonstrations, Share the Fun, county cooking and food show, and of course the county fair projects. Livestock and related projects continue to make up the largest project area with 36 percent of 4-H members exhibiting livestock.

Above: Swimming in a concrete pool was one of the highlights of 4-H'ers attending camp in Marshall County in the 1940s. Other activities included knot-tying and crop identification for the boys, and sewing buttonholes and keeping records for the girls. *Photo courtesy of the Marshall County Extension office.*

Left: Today's 4-H campers still love the pool, but workshops now have more focus on teamwork and communication, like this boat-building activity at Tri-County 4-H Camp in 2007. However, practical knowledge, such as wildlife education and safety, are still taught at 4-H camps today. *Photo courtesy of the Marshall County Extension office.*

Marshall County

Marshall County

Top left: Teen leadership activities keep older 4-H members excited and involved in 4-H, and gives them the opportunity for more in-depth problem-solving, decision-making, and teamwork. Marshall County 4-H has begun the new tradition of an annual 4-H Teen Retreat, which involves a float trip down the Illinois River. *Photo courtesy of the Marshall County Extension office.*

Bottom left: Community service has been a mainstay in 4-H throughout its one hundred years in Oklahoma. Marshall County's community service projects have ranged from mattress-making in the twenties to community beautification in the twenty-first century. *Photo courtesy of the Marshall County Extension office.*

Many people credit 4-H for providing first-time experiences and trips, such as national trips to Washington, D.C., and Atlanta. Many Marshall County 4-H'ers can now say 4-H gave them their first ski trip experience. *Photo courtesy of the Marshall County Extension office.*

Mayes County

Contributed by Connie Guthrie, Betty Bergman, Kaye Tipton, Debbie Zumstein, and Daniel Guthrie

At a 1939 demonstration, Betty Bradford Bergman and Patty Sharp Carpenter, Mayes County 4-H members, gave a demonstration on setting the table at Bryan Chapel School. *Photo courtesy of Betty Bradford Bergman.*

In 1926, newly hired agents E. B. Hildebrand and Irene L. Roberts started building the Mayes County Extension program. They started by enrolling 298 boys and 299 girls. They formed twenty-one junior clubs and held monthly meetings. Clubs were formed for the project areas of corn, wheat, oats, grain sorghum, alfalfa, soybeans, cowpeas, peanuts, fruit, and cotton. Other clubs formed were dairy, beef, swine, poultry, farm engineering, canning, home gardens, nutrition, and clothing. The purpose of these clubs was to teach and demonstrate improved farming and home demonstration practices. (This paragraph is a summary of the 1926 and 1927 Narrative Reports of county agents E. B. Hildebrand and Irene L. Roberts.)

We currently have seven clubs that serve each of the communities in our county. We have leaders with three generations of 4-H in their families. Karen Odle Chidester and Debbie King Zumstein, whose parents were 4-H members, have served as leaders for their children's 4-H club. Mayes County 4-H has been involved in traditional projects such as livestock, crafts, cooking, and sewing, but has also enjoyed the postmark project. Mayes County hosted the National Postmark Fair in 1997 and had many winners in the state fair and National Postmark Fair.

Mayes County has a history of leadership with individual agents serving in the county for many years. Anna Lee Rouk was the home demonstration agent from 1954 to 1983 and Janet Kleeman from 1984 to 2006. County agents who have served have included Phil Kennedy from 1971 to 1990 and Stan Fimple from 1990 to 1997.

Past agents, as well as current agents Mike Rose and Belinda Pfeiffer, have provided a good foundation for the Mayes County 4-H Program.

Mayes County

At the 1954 fair exhibits, Kaye Tipton and her sister, Suzanne Tipton Pietz, were very involved in project work and showed about one hundred projects they entered in the Mayes County Fair. *Photo courtesy of Kaye Tipton.*

Share the Fun, 1978. L-R: Mayes County Home Economist Anna Lee Rouk, Patsy Willis, Debbie King, and Brenda Smith. *Photo courtesy of the Mayes County Extension office.*

In 1998, Daniel Guthrie won the State Key Club Scholarship in 1998 as a Pryor 4-H member. Daniel attended the Washington Citizenship trip and was active as a county officer. Daniel Guthrie received the scholarship from Dea Rash. *Photo courtesy of the state 4-H office.*

McClain County

Contributed by LuGlena Moore and Connie Wollenberg

The McClain County 4-H delegates that attended State 4-H Roundup held on the campus of Oklahoma A&M College in Stillwater on August 3 through 7, 1936. Adult sponsors were Home Demonstration Agent Ivy Parker and Assistant Agriculture Agent Edward E. Davis. *Photo courtesy of Mrs. Florence Sewell-Farmer.*

McClain County 4-H is, and has always been, a strong program. The strength of 4-H in McClain County is partly due to a large number of third- and fourth-generation 4-H families.

The McClain County Extension office was established in 1914 with Mr. H. Garland serving as the Extension agent. Ray Parker, retired state 4-H specialist, began his 4-H career in 1929 in McClain County. Parker's main projects were poultry and public speaking. Parker received one of the highest honors by being inducted into the 2006 National 4-H Hall of Fame.

The McClain County 4-H delegation at State 4-H Roundup in 1966. *Photo furnished by McClain County OSU Cooperative Extension.*

McClain County

In 2006, McClain County entered the era of new technology with the GEO Clovers; this group is forging the way in the development of GPS maps and information that help serve our county emergency agencies. In 2009, McClain County had 318 4-H members in seven community clubs including Blanchard, Byars, Dibble, Newcastle, Purcell, Washington, and Wayne with five traditional project clubs: shooting sports, rabbits, horse, dairy goat, and poultry. McClain County has come a long way from the 1914 Boys' Corn Club and Girls' Canning Club to the science of GIS and wind energy in 2009. McClain County will continue "To Make the Best Better."

Betty Walker inspects a canning display for a McClain County Fair exhibit. Home Extension Agent Mary Eva Johnson presented a workshop on fancy packing green beans to McClain County 4-H girls. *Photo furnished by McClain County OSU Cooperative Extension.*

Throughout the years, McClain County has had several 4-H members serve as state officers. Karen Kay Speer served as the 1983–1984 state 4-H reporter; Angela Vieux was the 1999–2000 state 4-H secretary; Stephanie Bowen was selected as state 4-H ambassador in 2006–2007 and state 4-H vice president in 2007–2008; and Jayme Shelton was selected in 1998–2000 as state 4-H ambassador. McClain County has had several state 4-H record book winners and winners of various 4-H competitions at Roundup.

Charles and the late Peggy Bacon were honored in 2001, receiving the 2001 State Honorary 4-H Award presented by the Honorable Oklahoma Governor Frank Keating. *Photo furnished by McClain County OSU Cooperative Extension.*

McCurtain County

Contributed by Cathleen Taylor and Brad Bain

The 1988–1989 county officers. Seated, L-R: Amy London, game leader, and Cherrie Johnson, song leader. Standing: Ryan Martin, reporter, and Brad Bain, vice president. *Photo courtesy of the McCurtain County Extension office.*

"To Make the Best Better" is the 4-H motto, and McCurtain County 4-H'ers have always worked to make their projects, clubs, and county better. McCurtain County 4-H began in 1916 with two county Extension agents: R. C. and Eunice Blocker.

As times changed, 4-H clubs had to change as well. Thomas Hodge, as 4-H agent, led many youth in livestock judging. The next recorded 4-H agent in McCurtain County was Steve York. York worked with 4-H youth for two years. During that time, main project areas included livestock projects, cooking, sewing, and crop production. Gary Schafer took over in 1975 until Larry Wiemers served from 1978 to 1980. In 1981, Sammy Coffman was in charge of the 4-H program. Coffman coached state winning livestock judging teams five out of six years from 1983 to 1989. Coffman left Oklahoma State University Extension in 1986 to pursue another career and there was a vacancy in the 4-H agent position until 1995. Heather Winn then joined the McCurtain County 4-H team. After Winn left McCurtain County, a veteran McCurtain County officer Brad Bain from Valliant took over the 4-H educator position. In 2008, Cathleen Taylor took over the 4-H educator position after Bain moved to the agriculture educator position.

McCurtain County has had two state 4-H officers. Gary Marshall from the Broken Bow area was state 4-H president in 1975–1976. Following Marshall on the State Officer team in 1978 was Alan Marshall, who served as the state Southeast District vice president from 1978 to 1979.

Today, McCurtain County has twenty-five 4-H clubs across the county with thirty-five volunteers serving more than 575 youth. McCurtain County 4-H has a rich past and a bright future.

McCurtain County

Above: The 1989 judging team. L-R: Klyne Hughes, Shannon Daves, Cherrie Johnson, Brad Bain, and Dwight Clardy. Standing: Steven Clardy, Jaycee Johnson, Brad Wooten, and leader Sammy Coffman. *Photo courtesy of the McCurtain County Extension office.*

Left: The McCurtain County 4-H board of directors. Standing, L-R: Bob Severn, Glenna Herndon, Sharon Bain, and Willa Mae Parsons. Seated: Sandra Storey and Cindy Farley. *Photo courtesy of the McCurtain County Extension office.*

McIntosh County

Contributed by Lacy Whisenhunt

The 4-H program has long been a staple for youth in McIntosh County. According to Bill Hooten, who was in 4-H from 1946 to the 1950s at Cathy School, "Everyone joined 4-H. The 4-H people came to the school and signed kids up. Their club made items to send to the Muskogee State Fair."

McIntosh County is a rural county. For many years, 4-H has offered the opportunity for members to show their livestock. In 1940, Bane Whisenhunt joined 4-H to show his pigs. According to Bane, "the fatter they were the better they were." Bertha Whisenhunt, his wife, was also in 4-H, but not to show animals. Bertha's main project area was sewing. Bertha was a part of the Central High 4-H Club near Texanna. The first thing she ever sewed was a headscarf to enter into the county fair. One thing Bertha says she will never forget was "a seed collection she made with several other girls who were members at Central High. We handled the jalapeño seeds so much that it burnt our hands and we spent the next day soaking them in sweet milk."

Bane and Bertha, like many other families in McIntosh County, made 4-H a family event. They have had four children, eleven grandchildren, and one great-grandchild become a member of McIntosh County 4-H.

McIntosh County has always offered a variety of project areas and some very unique and creative activities. In the 1980s, they offered a clown troop. Several members were trained by a professional clown and would dress up and visit nursing homes and teach workshops on clowning around. McIntosh County has also had members who achieved the highest awards possible. In 1983, Mike Watson, a member of the Midway 4-H Club, was a 4-H national project winner in public speaking. McIntosh County has had thirty-four Key Club members starting in 1951 with Barbra Davis. We have also had members serve on the State 4-H Officer team, such as Christina Mays in 2003.

The 4-H members show their love for helping others by helping with an event registration. *Photo courtesy of the McIntosh County Extension office.*

Murray County

Contributed by Debbie Sharp

The exact year 4-H work began in Murray County could not be determined. However, early records from the 1930s made reference to thirteen organized 4-H clubs with 316 members. These members canned 770 jars of food, had 446 poultry projects and ten dairy cattle projects, and 128 girls raised gardens on twenty-three acres.

In the 1930s, 4-H clubs were located in Joy, Iona, Palmer, Hickory, Davis, Rayford, Gilsonite, Drake, Nebo, Woodland, Oak Grove, and Koeller. In 1932, the county lost one club located in a mining district because the superintendent of the mines would not allow the people to have any livestock. Today, seventy-seven years later, only one of these communities has a school open.

The 1940 annual records show twenty-one 4-H clubs with 244 boys and 264 girls. Adult volunteers were known as coaches. This was the first year boys could stay in a dormitory during 4-H Roundup rather than in private homes.

The 1950s found clothing as the most popular project for Murray County girls with 293 enrolled in the project and 242 completing a project. Recreation schools were held at Turner Ranch and 4-H and FFA Field Days were held on Governor Roy J. Turner's ranch.

The 1937 4-H Roundup delegates from Murray County enjoy a picnic in Stillwater. *Photo courtesy of the Murray County Extension office.*

Murray County

The decades of the sixties and seventies were highlighted by a national winner in photography and the county fair entries continued to be very popular.

The eighties and nineties found a slow decline in 4-H enrollment but an increase in teen leadership with four state vice presidents coming from Murray County.

The new century finds Murray County 4-H enrollment on the rise and reorganized into project clubs with a monthly community meeting. This format of club work has strengthened the program and members are staying in the organization longer, building strong leaders, and "Making the Best Better."

Murray County

Top right: In 2005, 4-H members participate in a project titled "Roving Readers." Members reading as a part of a community service project are Seth Coffey and David Lee. *Photo courtesy of the Murray County Extension office.*

Bottom right: One hundred years later, 4-H members continue to assist each other with project work. Blair Gee, a teen 4-H leader, instructs junior member Bryson Rogers. *Photo courtesy of the Murray County Extension office.*

Top left: Murray County 4-H members learn how to make a headscarf and potholder in 1950. *Photo courtesy of the Murray County Extension office.*

Bottom left: Murray County 4-H members learn how to can jelly in 1996. R-L: Bradley Riddle, Becca Runyan, Nikki Lodes, and Kelby McKinley. *Photo courtesy of the Murray County Extension office.*

Muskogee County

Contributed by Rodney King and David Adams

Muskogee County 4-H membership is widely diverse, representing the differences in county demographics. Muskogee County is an interesting mix of geography, people, and events. Established at the confluence of the Arkansas, Verdigris, and Grand rivers, Muskogee has river bottom farmland, native grass prairie, and rocky mountainous areas. Major events throughout the year include the Azalea Festival, Muskogee Regional Junior Livestock Show, Summer Farmers Market, Christmas Garden of Lights, Renaissance Festival, and Symphony in the Park.

Muskogee County 4-H currently has fourteen traditional 4-H clubs with more than 500 members. The county boasts an urban 4-H program that began over thirty years ago to meet the demands of urban youth in Muskogee County. Urban programming has grown to support a full-time 4-H educator delivering a wide variety of science and technology programs.

Muskogee County 4-H has a long history of diverse 4-H project work supported by a large, active volunteer leaders association. Local historians have documented 4-H events and activities as early as 1907. The first Muskogee State Fair began in 1907 and provided a number of events for 4-H members. Livestock exhibition has been a very large project area in Muskogee County since the early 1900s. The Muskogee State Fair ended after a poorly attended event in 1994.

The building Muskogee County Extension is housed in was built in the 1940s as a WPA project and was named the 4-H Building on the Muskogee State Fairgrounds. The building actually has 4-H Clovers designed into the rock. The building was originally built to serve as a dormitory for both male and female 4-H members to utilize while participating in the Muskogee State Fair.

In the late 1940s, Ira J. Hollar served as an Extension agent in Muskogee County. He solicited financial support from the Muskogee Noon Lions Club and started the Muskogee Regional Livestock Show in 1945. The Muskogee Regional Livestock Show grew into what is today the largest regional livestock show in the state, supported by the Chamber of Commerce and many area businesses. Ira J. Hollar, a legendary figure in Oklahoma 4-H, later became the state 4-H program leader.

Muskogee County has had a host of famous alumni. Muskogee County 4-H alumni include U.S. Congressman Mike Synar and former news anchor Karen Keith of Channel 2 News in Tulsa. Muskogee has also had a number of distinguished educators over the years. Ira J. Hollar, legendary Oklahoma 4-H state program leader; Ron Justice, current state senator; and Charlie Lester, the first Pistol Pete at Oklahoma State University, all served as county agents in Muskogee County.

Muskogee County is rich with much tradition and excellence in 4-H. The program continues to target the changing needs of clientele while providing life lessons through educational programming. Extension in Muskogee County has been and continues to be the center of youth development programming in Muskogee County.

Noble County

Contributed by Andrea McCluskey, Judy Farabough, Maxine Tautfest, and Nedra Will

The Oklahoma 4-H Program has made a great impact in the lives of youth. Through participation and involvement in 4-H projects, they learned valuable life skills including time management, public speaking, organizational techniques, team-building, leadership, and citizenship skills. They not only accomplished this in their own club, county, state, and country, but also carried these skills into their adult lives.

In Noble County, the county fair was a highlight of the year. It was challenging for members to enter as many top-quality projects as possible. Exhibits included cooking, sewing, gardening, record keeping, artwork, entomology, livestock, and other projects. Little did they know how many times in life those lessons would be used. County fairs have grown from tarp-covered show pens to fan-cooled barns and exhibit buildings with central heat and air. The 4-H members, their leaders and volunteers, and Oklahoma State University Extension personnel are the "heart and soul" of county fairs. Since their program began, Noble County also held the traditional 4-H contests including dress revue and Share the Fun. The members participated in county 4-H camps and State 4-H Roundup. The Extension Service provided programs where everyone in the community had an opportunity to participate.

The 4-H program taught our children the value of completion and competition in life. Records needed to be kept, work had to be done, and life lessons were learned. It taught them that sometimes you were not first place, but that you could encourage others to achieve also. Noble County is proud of the opportunities youth were given through 4-H programs. Our 4-H programs in Noble County and throughout the state of Oklahoma are still in full swing, changing the lives of future generations.

Earl Duane Hawkins and his Reserved Champion Angus, "Butch," in 1948. *Photo from* The Miller, *Jefferson Public Schools, Jefferson, Oklahoma.*

Delores and Donna Cox with their Grand Champion Poland China and Durocs in 1948. *Photo from* The Miller, *Jefferson Public Schools, Jefferson, Oklahoma.*

Right: Eldwyn Bradley and his Duroc in 1948. *Photo from* The Miller, *Jefferson Public Schools, Jefferson, Oklahoma.*

Below: Noble County 4-H members who attended the 1930 State 4-H Roundup conference in Stillwater, Oklahoma, in 1930. The tent where the conference was held can be seen in the background. *Photograph courtesy of Catherine Robinson.*

Noble County

All five daughters of Mr. and Mrs. Bob Farabough, a former Noble County Extension agent, learned how to care for and show sheep. The 4-H program is a family project where the older children teach the younger ones. *Photo courtesy of Judy Farabough.*

Noble County 4-H members learned many skills at the county 4-H camp held each summer. This photo was taken in the 1970s. *Photo courtesy of Barbara Pemberton.*

Larry Don Dewey and his Grand Champion Shorthorn, "Shorty," in 1948. *Photo from* The Miller, *Jefferson Public Schools, Jefferson, Oklahoma.*

Nowata County

Contributed by Milla Watts and Edith (Blum) Collins

This group of Nowata County 4-H members completed a 4-H record book for the 2007–2008 year and were rewarded with a trip to Incredible Pizza in Tulsa. Front row, L-R: Tyler Kallenberger, Morgan Kitterman, Colton Cantrell, Samantha Johnson, and Myles Lively. Back row, L-R: Dylan Allison, Benjamin Haddox, Lisbeth Haddox, Wyatt Lively, and Billy Bryant. *Photo courtesy of the Nowata County Extension office.*

Nowata County 4-H clubs have been providing a foundation for young people for many years. Edith Collins, a former 4-H member who was a part of the 4-H program from 1950 to 1955, says that "4-H matures boys and girls." At this time, there were at least seven 4-H clubs in Nowata County including Delaware, Childers, Watova, Diamond Point, Alluwe, Lenapah, and Armstrong. Collins adds that "one of the highlights for the 4-H members was the 4-H Roundup trip, when they were bused to the Oklahoma State University Stillwater campus for a time of friends, food, fun, and of course, lots of learning."

Some of the 4-H projects drawing the most participation from a group of over 200 fully active 4-H members were sewing, dress revue, canning, flowers, gardening, judging, public speaking, demonstrations, health, nutrition, and cooking.

The 4-H program influenced members and helped them prepare for their futures. "Some of them married

Nowata County 4-H members (L-R) Micro Watts, Myles Lively, Milla Watts, Wyatt Lively, and Kilo Watts helped paint the fair building as a part of their citizenship project in June 2009. *Photo courtesy of the Nowata County Extension office.*

straight out of high school and were perfectly able to properly and efficiently run their households, making them stronger and more outstanding members of the community," said Collins.

The 4-H program helps members face their challenges head-on, and take control of their own lives. For past 4-H members, the program was a part of their development. The volunteers are a major part of shaping children through 4-H. For Collins, the fondest memory of 4-H was all of them. Collins says 4-H is a necessary program and she hopes it will continue for another one hundred years.

Nowata County 4-H members canned grape jelly and tomatoes and baked cookies at the food, nutrition, and health workshop in July 2009. L-R: Micro Watts, Milla Watts, Lisbeth Haddox and Morgan Kitterman. *Photo courtesy of the Nowata County Extension office.*

Okfuskee County

Contributed by Ron Vick, Jan Maples, and Daniel Neely

An Okfuskee County 4-H member displays his beef livestock project. *Photo courtesy of the Okfuskee County Extension office.*

In 2009, Okfuskee County had ten community clubs and three special project clubs including dog, horse, and shooting sports, with 540 4-H members and fifty Cloverbud members. These clubs meet after school, at night, and on Sunday afternoons. Individual 4-H projects range from ATV to rocketry, performing arts, and livestock.

In comparison, according to the Okfuskee County Farm and Home Program Report dated December 27, 1949, there were twenty-four 4-H clubs in Okfuskee County with 700 members. The clubs were organized in the schools with a coach for boys and girls at each club. At the beginning of the year, each club member was enrolled in two or more projects and was required to keep them throughout the year.

The 1949 report goes on to list the following 4-H club work objectives:
1. To train young men and women to do a better job of farming and homemaking.
2. To train young people to associate with other young people to learn from their experiences.
3. To give young people a better understanding of some of the problems they will face when they start out on their own.

In 1949, 4-H club members conducted "result demonstrations" through their home projects as well as demonstrations in various phases of agriculture and homemaking. Also, in that era various county organizations sponsored 4-H projects and events including a pond maintenance and fish identification school, an annual banquet for 4-H club girls, spring dairy shows, medical examinations to the health members, buying pigs for members, and meat identification and training schools.

Although the projects may have expanded through the last sixty years, Okfuskee County 4-H members are still following the 1949 objectives of "learning by doing," meeting other 4-H members from the state and nation, and gaining insight into decision-making and problem-solving.

Oklahoma County

Contributed by L. Gale Goodner

Even though Oklahoma County was chosen as the seat for our state's capitol and has always been one of the most urbanized areas of Oklahoma, it still has a rich 4-H history. Like other counties, 4-H clubs in Oklahoma County were first called corn or cotton clubs and, later, junior club organizations, which were under the management of the home demonstration clubs. By the early 1920s, there were thirty-three 4-H clubs meeting with more than 1,500 members.

Volunteer leaders were called coaching staff and these teachers, parents, and other community leaders trained 4-H club boys and girls in better living skills that offered them an overall improvement in life. What becomes obvious from looking back at the records left from previous generations of 4-H members is how adaptable the 4-H clubs were and how focused they became on helping their members as the times changed. For instance, in 1943, Oklahoma County's 4-H clubs moved quickly to help teach many of the young girls left on the farms to drive tractors, care for dairy herds, and perform hundreds of other tasks that had once been completed by fathers or older brothers who had now left

Even in 1928, 4-H boys and girls gathered at the capitol to visit once a year and made sure that our representatives knew how important 4-H was to the communities it served. *Photo courtesy of the Oklahoma County Extension office.*

to fight in World War II. As more of the mothers began working in the factories, 4-H clubs helped teach older siblings to care for younger sisters and brothers as well as to see to the ordinary tasks of helping run a household when both fathers and mothers are away.

During the 1960s and 1970s, the focus on topics such as food preservation, livestock, and crop cultivation began to give way towards science and arts. One factor has remained obvious throughout the history of 4-H in Oklahoma County—its overall role and mission in the lives of Oklahoma's young people. In its one hundred years of service to the people of Oklahoma County, 4-H has continued to teach the better living skills that lead to an improvement in life for all of us.

In the fall of 1928, the Oklahoma County 4-H Program held a School House Poultry Show at the rural school serving the eastern section of the county. *Photo courtesy of the Oklahoma County Extension office.*

In 1933, Oklahoma County 4-H clubs competed for first place for their 4-H booth during the Oklahoma County Free Fair. The booths included samples of projects members had created during the past year and had entered in various competitions for first-place ribbons and awards. *Photo courtesy of the Oklahoma County Extension office.*

Oklahoma County

In 1941, it was quite an adventure to travel to Stillwater for statewide events. This group of Oklahoma County 4-H girls took part in a road trip to participate in one of the several judging schools 4-H held throughout the year. *Photo courtesy of the Oklahoma County Extension office.*

Forty-seven 4-H members from Oklahoma County attended the 4-H club camp at Kickapoo Boy Scout Camp in 1935. *Photo courtesy of the Oklahoma County Extension office.*

Members who attended the 4-H camp at Kickapoo elected to sleep outside in the heat of July. The cot-like beds serving the camp might not have been comfortable, but they were kept neat. *Photo courtesy of the Oklahoma County Extension office.*

Okmulgee County

Contributed by Donna Dillsaver and Ruth Bogie Langraf

Oklahoma 4-H is always holding events for all to participate in. *Photo courtesy of the Okmulgee County Extension office.*

On November 1, 1908, one of the first Oklahoma State University County Extension agents, Thomas M. Jeffords, was appointed to Okmulgee County. The actual year 4-H first started in Okmulgee County is not known. It is estimated to have started sometime in 1909. The 4-H programs have been an important part of the activities for the young people in the county. The programs provided by the 4-H leaders throughout the years were aimed at teaching the young members life skills—skills that would be beneficial to them as adults on their own or starting their own families.

Okmulgee County has always had a good representation in winning contests and participation in all events. With the help of their 4-H leaders, many club members have won state and national awards over the years.

Okmulgee County has had a number of 4-H members qualify to attend the American Royal in Kansas City, Missouri. Members also qualified in many areas to attend the 4-H Roundup held in Stillwater on the Oklahoma State University campus. Members could earn a trip to Roundup by winning at the dress revue/style show for the girls and the appropriate dress contest for the boys. Speech contest winners also qualified for Roundup.

At the county level, agents and 4-H club leaders held lots of activities in which all 4-H club members could participate. Many fun picnics were held in the summer and other parties were held in the fall. Then there were camps, the fair, and the final event for the year: the county 4-H achievement banquet.

Okmulgee County

The goal of the 4-H program is to encourage young people to develop new skills in personal achievement, leadership, and citizenship. Many successful community and state leaders grew up participating in 4-H activities and one family may carry on the 4-H tradition for several generations. This would not be possible without the dedication of the Extension agents, 4-H club leaders, and volunteers who help "Make the Best Better."

Young Oklahoma ladies pose for a photo for dress revue. *Photo courtesy of the Okmulgee County Extension office.*

Osage County

Contributed by Rick Rexwinkle

Osage County 4-H campers at the Osage Hills State Park in 2005. *Photo courtesy of the Osage County Extension office.*

The Osage County 4-H Program started out as a corn club. As time passed, canning and beef clubs were established in Osage County. By the early 1920s, local 4-H clubs were chartered in Pawhuska, Fairfax, and Hominy. It was at this time that county and district fairs were established. The first Osage County Fair was held in 1925 in Pawhuska. Three different district fairs were established in the late twenties—Hominy, Fairfax, and Grainola. The district fairs were held in Hominy and Fairfax until 1947, and the Grainola District Fair continued through August of 1981.

Osage County

The spring livestock show in downtown Pawhuska. *Photo taken by Williams Studio in Pawhuska, Oklahoma.*

An Osage County fair display of champions at the Osage County fairgrounds in Pawhuska. *Photo taken by Williams Studio in Pawhuska, Oklahoma.*

Livestock judging has been an important tradition with numerous judging teams participating in field days and state, district, and national contests, such as the National Western in Denver and American Royal.

Although agriculture is king in Osage County, leadership has also been an integral part of the program. State 4-H officers included Tom Tate (state president), Debbie Pelton (state vice president), Amy Gillick (state vice president), and Lindsey Long (state song leader).

The Osage County 4-H Program is rich with tradition. The 4-H Junior Livestock Show is held each March and was established by the Osage County Cattlemen's Association. Early spring shows were held in downtown Pawhuska on the street just north of the Triangle Building and old Duncan Hotel. In 1927, the brick arena was built at the county fairgrounds. However, this arena hosted its last show in 2009. In 2010, the exhibitors will be showing their livestock in the newly completed 300-by-400-foot arena.

Another summer tradition is 4-H camp. Many generations of 4-H'ers have attended camp at Osage Hills since the park was established in the 1930s. Most years, it has been filled to capacity with members enjoying educational workshops and recreational activities.

Ottawa County

Contributed by Mary Clogston and Debbie Gaines

In 1957, Gerald Rendel and Gary Boyd gave a demonstration on livestock conservation. *Photo courtesy of the Ottawa County Extension office.*

Ottawa County is located in the northeast corner of Oklahoma. The county is well known for the beautiful Coleman Theater. Ottawa County is very proud of its American Indian heritage.

Ottawa County 4-H dates back to the time of statehood. The first 4-H county agent was Odes T. Pogue on January 18, 1915. Mr. and Mrs. B. E. Boyd lived in the Lone Star community since 1883, and their eleven children were 4-H members in the Lone Star 4-H Club, which was the first club in the county. Records show a combination of 202 years in club work and eighty-six years in home demonstration work. The Boyd family has great-grandchildren who are still active in Ottawa County 4-H today.

Ottawa County

Lila and Lester Boyd (the son of B. E. Boyd) met each other when he was county president and she was county secretary. Lila was outstanding poultry winner in 1939 and district leadership winner the same year. She was a member of the state canning judging team and won two trips to National Congress. Lester showed the champion Angus in 1939 at Tulsa State Fair and was high individual at Joplin Fair in livestock judging.

Lila (Wilson) Boyd holds the trophy she won in 1939 in the State Poultry Contest. *Photo courtesy of the Ottawa County Extension office.*

L-R: Richard Schubert, Randy Gardner, Mayor Wayne Pack, and Alvin Krumley constructed new county 4-H signs at each end of the county lines. *Photo courtesy of the Ottawa County Extension office.*

In 1933, Ella Lou Reynolds was eight years old when she joined the Victory 4-H Club. Ella Lou's fondest memories were county rallies, style shows, broiler shows, and fairs. After graduation, she became a schoolteacher

The whole Boyd family belonged to the Lone Star 4-H Club in Ottawa County. Back, L-R: Johnny, Wayne, Bill, Kenneth, Clinton, Lester, and James Boyd. Front, L-R: Frances (Boyd) Simmons, Helen Boyd Rowley, father and leader Bert Boyd, mother and leader Macy Boyd, Bessie (Boyd) Gibson, and Juanita (Boyd) Hayes. *Photo courtesy of the Ottawa County Extension office.*

and the Fairland 4-H Club leader for forty-four years and received the Outstanding Club Award in 1966.

Sharing the Outstanding Club Award in 1966, JoNell Babst, who was an OSU Extension educator in 1988, is recognized for her two children, who were state and national entomology winners and Hall of Fame recipients. JoNell says that "after God and family, 4-H is her next love."

In 1999, Jimmy Rexwinkle was recognized for many awards on district and state levels. Offices held were Northeast District president, state secretary, and state president. In 2000, Jimmy joined the state Hall of Fame.

Currently, the Ottawa County 4-H Program consists of seven clubs and six project clubs that serve approximately 443 members. Working over fourteen civic projects a year as a group, community service is a major part of Ottawa County's program.

Lester and Lila Boyd were recognized at State Senior Roundup for the National 4-H Alumni Award. Pictured with the couple is Myra Whitehead, Ottawa County Home Extension, 1950. *Photo courtesy of the Ottawa County Extension office.*

Ottawa County

Left: Lila Boyd from Ottawa County attended the 1939 Oklahoma delegation to National 4-H Congress. *Photo courtesy of the Ottawa County Extension office.*

Below: Larry Boyd, in the front, holds Poppy. All the calves were won by Raymond Boyd. Raymond Boyd, holding the middle calf, won a heifer for winning the milking contest at the Tulsa State Fair. He named her Noble. Clinton Boyd, in the back, is holding Raymond's heifer he won at the Louis Chanenworth Dairy Contest. *Photo courtesy of the Ottawa County Extension office.*

Ottawa County delegates fly to State Senior Roundup in Stillwater. *Photo courtesy of the Ottawa County Extension office.*

Pawnee County

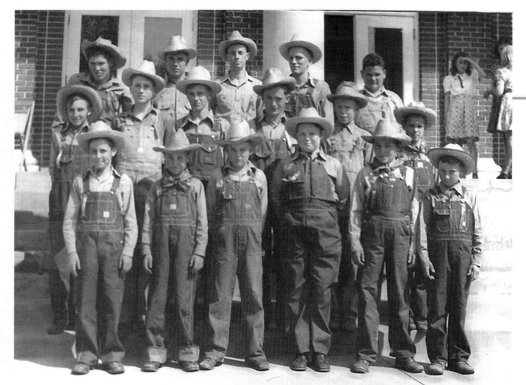

In 1942, these boys were entered in the appropriate dress contest. Notice the overalls they wore for the casual wear. *Photo courtesy of the Pawnee County Extension office.*

After looking back through historical information and visiting with past and present 4-H members, volunteers, and leaders, many things have changed, but many things have stayed the same.

Project areas in the past were focused on things needed for basic living. Projects were livestock, gardening, sewing, and food preservation and preparation. Today, these projects are still around, but newer projects are being focused on things like leadership and community service. Two projects Pawnee County 4-H members have shown a great interest in over the years are shooting sports and equine. This has led to two 4-H project clubs: Pawnee County Shooting Sports and Pawnee County Horse Club.

Camp and 4-H leadership conferences have always been popular events with Pawnee County 4-H. At one time, Pawnee County 4-H members attended camp at the Methodist Church campgrounds in Pawnee and later went to Lew Wentz Camp in Ponca City until they became a unit with Osage County. As a unit, they joined Osage County to attend camp at Osage Hills State Park. Today they are no longer a unit, but continue to go to 4-H camp at Osage Hills State Park with Osage and Okfuskee counties. For years, Pawnee County attended State 4-H Roundup in Stillwater, Oklahoma, on the Oklahoma State University campus.
In earlier years, members competed in speech contests, dress revues, demonstrations, and other project areas in competition with other members around the state. Today, members attend State 4-H Roundup for some of the same reasons, but more for the educational workshops, leadership skills, and the naming of the state 4-H ambassadors and state 4-H officers. Other 4-H leadership conferences and camps that Pawnee County 4-H members have attended in the past and present are the 4-H Citizenship Washington Focus trip, National 4-H Conference, Kansas City Global Conference, Northeast District Youth in Action Conference, and Kee Gee Camp.

Pawnee County

At the end of every year, Pawnee County 4-H has an achievement banquet. The banquet is to recognize 4-H members on their achievements for projects they have listed in their medal folders. They also are recognized for the trips and workshops they have attended. Pawnee County 4-H awards three outstanding members each year with the Mable Brien Award, Edna Sallee Award, and 4-H Hall of Fame. During this time, they recognize the local 4-H club leaders.

Above: In 1949, these 4-H Livestock members listen to Mr. W. J. Beck, associate livestock specialist, explain some points in "showing" a beef calf. *Photo courtesy of the Pawnee County Extension office.*

Right: Lone Jack 4-H Club members work on club projects at their meeting. *Photo courtesy of the Pawnee County Extension office.*

These delegates from Pawnee County attended 4-H Roundup in Stillwater, Oklahoma, in May 1957. *Photo courtesy of the Pawnee County Extension office.*

The delegates from Pawnee County who attended State 4-H Roundup in Stillwater, Oklahoma, on the Oklahoma State University campus in May 2007 were Tracie Hoeltzel, Maverick Thurber, Elizabeth Harrelson, Brandon Keys, James McAlister, Katsy Funkhouser, Amy Presley, Torie Funkhouser, and Lauryn Funkhouser. *Photo courtesy of the Pawnee County Extension office.*

The 4-H program has been an important part of Pawnee County and the families within the county for many years. Numerous families have come and gone through the years, but with the support of the Oklahoma Cooperative Extension Service, along with volunteers and working with people within the community, the 4-H program will continue "To Make the Best Better."

In the summer of 2007, junior 4-H members took an educational trip to tour the Blue Bell Creamery. *Photo courtesy of the Pawnee County Extension office.*

Payne County

Contributed by Bonnie Fisher Muegge, Dea Rash, and Brett Morris

Bonnie Fisher, a member of Hillside 4-H in Cushing, exhibited the Grand Champion Fat Lamb at the Payne County Junior Fatstock Show in 1953. *Photo courtesy of the Payne County Extension office.*

The Payne County 4-H Program has been instrumental in the lives of many for the past one hundred years. Early 4-H work focused on agriculture projects, such as sheep, beef, swine, gardening, food preparation, clothing, and food preservation.

Today, 4-H members still do traditional projects, but they also are involved in new projects such as shooting sports, dog care, and meat goats. Public speaking, citizenship, community service, leadership, and achievement projects have stood the test of time and have taught valuable life skills for a century.

For many years, Payne County 4-H members went to 4-H camp each summer at Lew Wentz Camp in Ponca City. In recent years, they have attended 4-H camp at Saints Grove Camp east of Stillwater.

The 4-H members attended State 4-H Roundup each summer, participating in competitive events and officer elections. In the 1950s, Oklahoma A&M College and Southwestern Bell sponsored a State 4-H Personality Improvement Program, a project no longer offered today, designed to develop all aspects of the individual. In 1957, Bonnie Fisher from Hillside 4-H in Cushing was named state winner in this competition. The 4-H members still attend State 4-H Roundup today, where one of the current highlights is the naming of state 4-H officers and ambassadors. State 4-H ambassadors from Payne County include Monica Combrink, 2000–2001; Steven Anderson, 2000–2001; Jamie Weber, 2001–2003; Katy Selk, 2004–2007; Alea Sharp, 2004–2007; Kendra Rash, 2008–2010; and Madison Rash, 2009–2010.

Payne County 4-H has also had several state 4-H officers including Roberta Smith, secretary, 1952; LaRae Dowell, secretary, 1954; Brenda Adams, secretary, 1955; Lana Grooms, song leader, 1961; JanEtte Dahms, reporter, 1970; Tammy Retherford, song leader, 1977; Anna Williams, vice president, 1993; Heather Williams, vice president, 1994; and Erin Reece, vice president, 1999.

Payne County 4-H members have worked on county and state record books for many years. Record books began with handwritten information on report forms and "penny pages" with many pictures and newspaper clippings to handwritten project work on participation forms with project sections

Payne County

Bonnie Fisher also exhibited the Grand Champion Fat Lamb at the Payne County Fair in 1949. *Photo courtesy of the Payne County Extension office.*

Sixteen Hillside 4-H Club members from Cushing made dolls for the Seventeen Magazine Doll Contest. *Photo courtesy of the Payne County Extension office.*

Gathering nylon hose to send to Korea was a citizenship project of Hillside 4-H Club in Cushing. *Photo courtesy of the Payne County Extension office.*

with one page of pictures and without newspaper clippings. In 2008, members began typing county record books on computer-generated forms. Through the years, many Payne County 4-H members have been named state 4-H project winners with their state record books with one or more winners nearly every year.

Several Payne County 4-H members have been inducted into the state 4-H Hall of Fame: Gordon Dowell, 1952; Brenda Adams, 1956; Bonnie Fisher, 1958; Nancy Stiles, 1966; Tamara Morrison, 1974; Kent McVey, 1980; Jill Anderson, 1996; and Steven Anderson, 2000.

For many Payne County families, 4-H and the Cooperative Extension Service has been an integral part of their lives for many generations and the Payne County 4-H Program has helped 4-H members "To Make the Best Better" for a century.

Bonnie Fisher (right) used her sewing skills that she learned in 4-H to help other girls with their sewing projects. *Photo courtesy of the Payne County Extension office.*

A group of modern-day Payne County 4-H members are shown after competing in the Northeast District Public Speaking Contest. Public speaking has always been important to Payne County 4-H'ers, who are always a threat to win in any public speaking competition. *Photo courtesy of the Payne County Extension office.*

Pittsburg County

Contributed by Greg Owen

The Pittsburg County 4-H Program is probably best known for two of our most famous alumni: Reba McEntire, a former member of the Kiowa 4-H Club, and Carl Albert of Bugtussle, Oklahoma. Pittsburg County 4-H has a strong history. Currently, the county's program has one of the largest enrollments in the state and has for the past several years. Charles Smith has been the one and only state 4-H president and served in 1951. Smith currently owns Canadian Valley Telephone Company. The oldest records found of 4-H work in the county include a secretary book by Haileyville 4-H Club from 1929 and a county fair book from 1933.

Some past and current 4-H educators on record include Ron Vick, Val Terry, Larry Wiemers, and Greg Owen. Pittsburg County has one of the largest county fair programs in the state with an average of 1,500 non-animal entries and 2,500 total entries each year. Pittsburg County's 4-H program has had sixteen Oklahoma state 4-H ambassadors, 123 Oklahoma 4-H Key Club members, and ten state 4-H officers. Recently, county support has added a second 4-H educator to our program named Mike Carter.

Currently, there are approximately thirty-five 4-H clubs with an average annual enrollment of more than 1,000 members. School enrichment is a big program for our county where we reach an excess of 5,000 students annually. Offices were held in the Pittsburg County Courthouse until quite recently; sales tax revenues helped build a new Extension office that opened in 2008. Pittsburg County has a weekly radio program on KNED and two weekly columns in the *McAlester News Capital* and the *Bargain Journal*.

Pittsburg County 4-H has a strong and storied history, but the future looks even brighter for this wonderful county program.

Pontotoc County

Contributed by Pontotoc County Extension

Asa Hutchison, an Ada businessman, presents Jerald Barton of the Fitzhugh 4-H Club with the achievement trophy. The award was presented annually at the Pontotoc County Fair to the child totaling the highest points based on placings at the fair. *Photo courtesy of Pontotoc County 4-H.*

Pontotoc County 4-H began with guidance from the father of Extension work in Oklahoma: W. D. Bentley. Pontotoc County has since grown to become a diversified program with 500 4-H members involved in science and technology projects and traditional agricultural projects. There are seventeen junior and senior chartered school or community clubs and five project clubs.

Rich traditions in Pontotoc County mark its place in history. Pontotoc County raises funds through the 4-H Royalty Contest. In twelve years, members and volunteers have raised $166,503 to support local and county program efforts.

The Asa and Marjorie Hutchison Award is one of the most sought-after honors. It was established in the 1950s by Asa Hutchison to encourage participation in the county fair. One boy and one girl were chosen. Today, the top two members vying for this award are chosen regardless of gender.

Public speaking is one of the most popular projects in Pontotoc County. Two speech contests are held each year. The Fall Public Speaking Contest incorporates public speaking, posters, and technology while the annual County Roundup is specifically public speaking. County Roundup dates back to the 1940s.

Above: Linda Jean Kirkley demonstrates a trick to develop good posture as teammate Patty Lamirand tells how in the health demonstration given at the 1957 County 4-H Roundup. Linda and Patty are members of the Pickett 4-H Club. *Photo courtesy of Pontotoc County 4-H.*

Left: Adam Daniel of the Homeschool 4-H Club gives an illustrated presentation about video games during the 2009 county 4-H Roundup. *Photo courtesy of Pontotoc County 4-H.*

Pontotoc County

Since 1959, four members have served as state 4-H president and four members have earned membership in the state 4-H Hall of Fame.

Life skill development is an important goal of Pontotoc County 4-H. One 4-H member summed it up best when she said, "I like being in 4-H because it gives me the opportunity to make a difference in the world just one small step at a time. It also allows me to be a positive role model for anyone who looks up to me."

The 2008–2009 4-H Royalty who served during the centennial year are, L-R, Escort Zachariah Smith, Homeschool; Queen Amanda Wilson, Homeschool; Princess Angel Wade, Meat Science; and Escort Rowdy Veal, Meat Science. *Photo courtesy of Pontotoc County 4-H.*

Pottawatomie County

Contributed by Sarah Weeks and Lindsey Hix

While some would consider thirteen an unlucky number, the number fourteen is a badge of pride for Pottawatomie County 4-H. Pottawatomie County has had fourteen youth serve as an Oklahoma 4-H state officer. The 4-H program is a strong family tradition in Pottawatomie County. In addition to leadership, Pottawatomie County 4-H has also established some traditions in animal science. In 2009, eleven youth participated in the Oklahoma Youth Expo as the third generation of their family to exhibit livestock under the name of Pottawatomie County 4-H. Many Pottawatomie County youth are second- and third-generation 4-H members. They have a true love and understanding of the concepts behind Head, Heart, Health, Hands, and Home.

Located forty-five minutes east of Oklahoma City, Pottawatomie County is unique in that over half its population resides in only one of the twelve towns inside county lines. Pottawatomie County produces substantial quantities of beef, wheat, and hay. Currently, there are seventeen 4-H clubs in Pottawatomie County. Most are community clubs. However, there are also active shooting sports, sewing, and horse and dog clubs as a part of the 4-H experience in Pottawatomie County.

Michael Larson, now a thirteen-year-old Pottawatomie County 4-H officer, meets his first show heifer. With Michael is his grandfather, Jesse McGaha. Five generations of the McGaha family have shown Hereford cattle in Pottawatomie County. *Photo courtesy of the Pottawatomie County Extension office.*

In 1998, through the guidance and support of then County Extension Director Don Britton, Pottawatomie County Extension began to receive funding from the county's one-cent sales tax. This allowed Pottawatomie County to fill two full-time 4-H educator positions, and provide more support to 4-H and youth development in Pottawatomie County. In 2005, Pottawatomie County's OSU Extension Service moved into its permanent home on the corner of Acme and MacArthur in Shawnee.

Pottawatomie County

I HAVE BEEN, I AM, AND I WILL BE,

A LEADER FOR YOU

ELECT

FAYE STONE
CENTRAL DISTRICT VICE PRESIDENT

★ *National Citizenship Short Course Delegate 1972*
★ *Two Years Local Vice President*

Faye Stone, 2009 Pottawatomie County 4-H Lifetime Leader of the Year, ran for Central District 4-H vice president in the early 1970s. Faye currently has children and grandchildren in Pottawatomie County 4-H. *Photo courtesy of the Pottawatomie County Extension office.*

A group of 4-H members gather outside the Pottawatomie County OSU Extension office in the fall of 2008. *Photo courtesy of the Pottawatomie County Extension office.*

Pushmataha County

The Pushmataha County 4-H Club had 452 members for the 2008–2009 year. All seven county schools have a local club and the school administrations strongly support 4-H activities, ranging from speeches and demonstrations to leadership conferences and livestock activities. Project clubs also flourish in the county. The monthly shooting sports club meeting is eagerly anticipated. Likewise, the goat club meets monthly to provide information and fun activities to county youth.

The 4-H program is a longstanding and multi-generational tradition in Pushmataha County. Many families now have been involved in 4-H for three generations. Now we see former 4-H members at county events helping their children and grandchildren with the same type of project or exhibit they worked on as a youth, such as sewing, making jelly, or livestock projects.

Pushmataha County 4-H'ers participate in several countywide events each year.

The county Junior Leadership Conference is held annually at the Tuskahoma School. Geared toward members aged nine to twelve years, participants attend workshops on leadership activities and work on projects for the county fair. Older members develop leadership skills as they conduct or assist with these workshops.

The county fair always brings excitement as 4-H members and adults enter their baked goods, artwork, sewing, horticulture, and livestock exhibits. First-place 4-H exhibits are taken to a higher level of competition at the Tulsa State Fair, and a significant number will win or place highly there.

Artistic talents are displayed at the pumpkin festival. The decorations may be carved, painted, or both, and include storybook characters, spooky images, or objects from the wildest imaginations.

More leadership and fun activities are found at the county junior lock-in, held annually at the Moyers School. Workshops may range from "First Aid" to "Record Books are Fun."

Animal exhibits take front row at the junior livestock show each March. Cattle, hogs, sheep, and goats are exhibited for both financial incentives and bragging rights. Parents and grandparents are sure to be there to see the outcome.

The county rally and speech and demonstration contest allows individuals and teams to conquer their nerves as they compete for prizes and campaign for countywide 4-H offices.

Share the Fun is always enjoyed by both the participants and the audience. Performances may range from skits to songs to tap dance.

Each school year concludes with the end-of-year bash, a celebration of the successes of the previous year, and recognition of award winners and notable volunteers.

Each year, one or more 4-H members from our county are selected as county Hall of Fame members and their pictures are added to our "Wall of Fame." This tradition began about forty years ago. Many of the people pictured on this wall attended various colleges and universities on scholarships for academic excellence or talent as a judge of livestock, and credit their 4-H experiences for starting them along their chosen path. These outstanding 4-H members have excelled in later endeavors as well, with some achieving local, state, and even national recognition as doctors, teachers, business owners, bankers, one Oklahoma State University professor, and one state senator. Several have served as adult volunteers for local 4-H clubs, giving back to the organization that meant so much to them.

One of those is former Wall of Fame honoree and current 4-H Adult Volunteer Mona (Baggs) Dennis, who received a token of appreciation for her years of service to the Tuskahoma and Clayton 4-H clubs at the 2008 end-of-year bash.

The 4-H members are also involved in numerous community service projects. Groups of 4-H members worked with other community groups to clean up Kiamichi Park. Other projects include the annual Clean Up Antlers, a Canned Food project each holiday season, and a coat drive.

Pushmataha County 4-H has come far from the early days of livestock, horticulture, crops, sewing, and cooking, but as early members have in the past, they continue to give back time, effort, and financial support to help expand the role of 4-H in the lives of Pushmataha County youth.

Roger Mills County

Contributed by Dale and Judy Tracy

Above: Roger Mills County 4-H members Nels Olson, Mike Smith, Chris Minor, and County Agent Dirk Webb won the 4-H National Championship in Pasture and Range Judging at the 29th International Pasture and Range Judging Contest in 1980. Olson and Smith were first- and second-place individuals. *Taken from* The Cheyenne Star *newspaper*.

Left: Klina Potter in 1915. Klina Potter started the first 4-H clubs known as Pig Clubs in Roger Mills County. She organized a chicken banquet to raise money to hire a superintendent, later known as a county agent. *Photo provided by Dale and Judy Tracy*.

In 1917, Klina Potter was teaching at the Fairview one-room school near Durham, Oklahoma, located in the northern part of Roger Mills County. She encouraged her gentlemen scholars to wash and show a pig and encouraged the lady scholars to can tomatoes. The first 4-H clubs were known as pig clubs. Klina organized a chicken banquet to raise money so the Pig Club could hire a superintendent, later known as a county agent. Soon the county superintendent of schools, T. C. Moore, organized boys' and girls' agricultural clubs throughout the county. The first county agent was Mr. Rathbun.

Klina married John Casady in 1919. He was already editor of the *Roger Mills Sentinel*, and later the editor of the local newspaper, *The Cheyenne Star*. Klina Casady continued to support the 4-H clubs and was named as a state 4-H honorary member in 1949. After this, she began to award a silver dollar to the first-place record books in each year of 4-H work. This is continued today as the Klina Casady Award.

Roger Mills County has grown to ten 4-H clubs with over 154 members and nineteen volunteer leaders. There have been three state 4-H presidents: Edgar

Roger Mills County

Roger Mills County 4-H members and leaders at Roundup in Stillwater, 1932. *Taken from* The Cheyenne Star *newspaper*.

McVicker, 1934–1935; Keith Tracy, 1989–1990; and Shannon Ferrell, 1994–1995. The Cheyenne Senior 4-H Club became state scholarship donors in 1988 and continued until 2006.

Randomly selected national honor winners included Nels Olson, beef; Brad Cowan, Santa Fe; Nicky Smith, agriculture; and Keith Tracy, photography.

State 4-H honorary members were Clara Davis and L. L. and Lorena Males. State 4-H Volunteer of the Year Award winners were Judy Tracy and Betsy James.

Roger Mills County 4-H members and leaders attending 4-H Roundup in Stillwater, 1957–1958. *Taken from* The Cheyenne Star *newspaper*.

Rogers County

Contributed by Donna Patterson

A group of Rogers County 4-H members from the earlier days of 4-H. *Photo courtesy of the Rogers County Extension office*.

The 4-H program has been a "family tradition" in Rogers County since its beginning. Rogers County, which is located in the northeast part of Oklahoma, is a rural community with a strong background in agriculture and crop production. The earliest record of organized 4-H clubs in Rogers County is about 1916, with those clubs starting out as demonstration clubs. The 4-H clubs were a way for youth in the rural communities to socialize and meet new people. It gave many of them a sense of community instead of isolation in rural areas.

Some Rogers County families with a rich 4-H history include the Blakley, Collins, Dorsey, and Guilfoyle families. These families saw the importance of 4-H in teaching responsibility, leadership, and citizenship. Many of these families are now working on a fourth generation of 4-H'ers.

Rogers County has quickly grown into a busy urban county. However, it still holds that strong practice of agriculture production. Currently, the Rogers County Cooperative Extension Service oversees ten clubs with 315 4-H youth. It also serves over 600 youth through school enrichment programs. While staying true to our beginnings in agriculture, home economics,

Rogers County

and community service, we have embraced the future of science and technology. The 4-H members are enrolled in fifty-five different project areas with the top five projects including leadership, citizenship, horse, shooting sports, and beef. Rogers County holds true to the 4-H motto of "Learning by Doing." Whether it is planting flowers, exhibiting at the county fair, or using GPS to map storm drains, Rogers County 4-H will continue "To Make the Best Better."

Today's 4-H community service projects span many activities, including mapping storm drains. *Photo courtesy of the Rogers County Extension office.*

Seminole County

New Lima 4-H members Brenda and Cindy Kifer. *Photo courtesy of the Seminole County Extension office.*

Although Seminole County 4-H began early in the twentieth century, it has been impossible to pinpoint an exact date. It has been extremely active during some periods and rather inactive in others, but 4-H has remained intact through both good and bad times.

During war times and times of natural disasters, members came out in full force in the areas of safety and emergency preparedness with as many as 1,185 students participating. Personal development and recreation leadership were also strong with about 1,200 members participating. Also with about 1,200 members participating were food and nutrition, art and crafts, and surprisingly, government and citizenship programs.

Seminole County

Some figures show that during the early days of Seminole County 4-H, there were as many as 20,308 members enrolled in project activities. A little more than half were boys, but not many more. In all project areas, the ratio of boy to girls was fairly even.

Newspaper articles and personal stories from past members indicate that 4-H touched the lives of both members and people with whom members were involved. Many stories were inspirational, telling of how 4-H gave many students a desire to succeed in future endeavors as well as just fit in with fellow classmates.

The community has always been very involved with all aspects of Seminole County 4-H. The Kiwanis Club and the Lions Club have been particularly involved with 4-H events. They have been very generous with their time as well as with their finances.

Some activities that county 4-H'ers have always been active in are the county dress revue, county rally, and Roundup. The members also greatly anticipate 4-H Day at the Capitol.

This is a small summary of our activities over the years, but hopefully an inspiration for more to come.

Top left: In 1943, Null and Robertson of Seminole were the first-place champions of livestock conservation demonstration. *Photo courtesy of the Seminole County Extension office.*

Left: In this fiftieth anniversary of 4-H picture taken in 1959, Lincoln and Seminole were the state champion teams. *Photo courtesy of the Seminole County Extension office.*

Sequoyah County

Contributed by Donna Jamison, Watie Goodwin, and Norma Goodwin

McKey grade-school 4-H members. *Photo courtesy of the Sequoyah County Extension office.*

Spend a day in Sequoyah County and you will experience 4-H as a family affair. Family is the tower of strength behind every 4-H member. The encouragement and assistance of parents or a good leader helps encourage strong skills in Sequoyah County youth as they grow up. No matter what the event is, parents and grandparents both attend the events of their children and grandchildren. Yes, 4-H has changed in many ways over the years, but the basics are still the same.

The many skills the children learn range from how to conduct a meeting to preparing for a 4-H speech and demonstration. Past Sequoyah County 4-H'ers have learned a vast amount of education, leadership, and citizenship skills that have transferred to and are applied in adulthood.

Donna Jamison, a former 4-H leader, said, "One of the highlights of our 4-H club experience was when all three of my children were winners of the salad-making contest at Muskogee State Fair."

Delanna was a first-year member, Kendall a junior, and Ladonna a senior. To this day, the family enjoys the recipes they used for their salads. "Many new projects for the 4-H program were developed, adding new spark, but the pledge continued to keep our heart in our community, country, and world," Jamison said. "Also one of the projects our club did was Adopt-A-Highway. If you have picked up a mile of trash, you will think twice before throwing something out of your car window or putting it in the back of your truck bed." As a former 4-H leader, "To Make the Best Better" is how Jamison continues her life!

Watie Goodwin, a former agricultural Extension educator, served not only Sequoyah County, but also five other counties: Adair, Muskogee, Delaware, Cherokee, and Mayes. He had to arrange time to service all six counties and was spread very thin to accomplish his task as a leader for the clubs. Goodwin said that he "was very proud of the children in 4-H clubs in all six counties along with his own four children." He was proud even when other children won awards over his own children. The children winning the awards made him feel like he was

being a good leader and he was helping to teach the children to grow through the program.

As a family, the Goodwins both were involved in 4-H. His wife, Norma, had a club of all girls. She took them to the district's Share the Fun along with many other events. Their own family was very involved in 4-H from the boys showing steers in the livestock shows to their daughter winning awards in Share the Fun, along with the county fair events. Watie said, "I didn't play favoritism with my 4-H children along with my own four children." He added, "Life as a 4-H leader was very fulfilling and I am proud of what I did."

The 4-H program influenced the lives of three classmates from McKey. L-R: Thelma Covington Dodd, Glenn "Cat" Taylor, and Ora Lee Dobbs Kirk. *Photo courtesy of the Sequoyah County Extension office.*

Stephens County

Stephens County was formed in 1907 from part of territorial Comanche County and part of the former Chickasaw Nation. Containing 891 square miles of land, it is twenty-seven miles wide and thirty-three miles long with the county seat of Duncan located in the middle of the county. At the time of statehood, Stephens County was mostly cattle country and for many years was home to one of this state's largest and most successful county fairs. By 1918, Stephens County was quickly shifting from ranching and farming to oil production.

In May of 1909, Stephens County was fortunate to have their first county agent, W. N. McPherson, working with agricultural producers in the

Above: A group of Stephens County girls perform a routine at Share the Fun. *Photo courtesy of the Stephens County Extension office.*

Left: A group of Stephens County 4-H members. *Photo courtesy of the Stephens County Extension office.*

county, and in 1918, A. G. Bowles was hired to work with the 4-H youth of Stephens County. Esther Martin was hired in the summer of 1922 as the first Stephens County home demonstration agent to assist the ladies of the community. Over the years, the county Extension office has been housed in the U.S. post office, in the courthouse in the Main Street square, in the basement of the current courthouse, and in 2007 was moved to the Stephens County Fairgrounds.

Over the years, Stephens County 4-H has been strong in both agriculture and family and consumer science projects. While almost all of the clubs have been school-based and connect to the community through the school system, there have also been numerous project clubs. Today, they continue to carry on the tradition with project clubs for horse, livestock, and shooting sports. Stephens County 4-H has been home to many outstanding county and state alumni including one international champion in trap shooting and Seoul Olympic competitor.

The 4-H program has encouraged youth to research different methods of agricultural practices. *Photo courtesy of the Stephens County Extension office.*

Stephens County poultry exhibits.

Texas County

Contributed by Arleen James, Lewis and Anna Mayer, Joan Chuesburg, Loretta Reeves, Dolores Piepho, Pat Vaughn, Dr. Kenneth Woodward, Kaye Tipton, the Margaret Teel Estate, and Kaly Grunewald

The Texas County 4-H Program has had a rich history of state and national 4-H winners since the beginning of the program. Oklahoma 4-H Hall of Fame winners from Texas County include Renetta Reeves, Jimmy Quinn, Larry Quinn, Garvin Quinn, Pete Teel, and Jerry Harke.

Jimmy Quinn and Loretta Bauer Reeves shattered tradition by being the first boy and girl from the same county to be simultaneously selected for National Club Conference in Washington, D.C. Robert Sheets, a county agent, stated, "For a

Above: Texas County 4-H members in their official dress in the early 1940s era. *Photo courtesy of the Texas County Extension office.*

Left: The Loyal Doers Club still continues to be an active club even many years after it was established. Each year, the club actively participates in kids helping kids. *Photo courtesy of the Texas County Extension office.*

Texas County

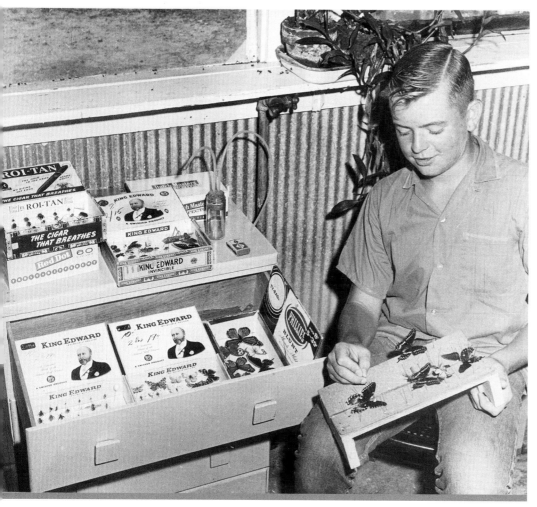

Above: A young Texas County member works on his entomology project. *Photo courtesy of the Texas County Extension office.*

Pottery is a unique project area that Oklahoma 4-H members enjoy. These Texas County members enjoy making pots. *Photo courtesy of the Texas County Extension office.*

county to have two conference delegates in one year leaves you somewhat speechless" (*Guymon Daily Herald*, 1957).

In 1915, the first county agent was B. M. Jackson, and the first home demonstration agent was Lula Jackson. Current Extension educators include Steve Kraich and Arleen James.

In 1929, the first annual Panhandle 4-H Club Roundup was held. In the thirties and forties, the girls' and boys' 4-H club had over 1,500 members. The first achievement banquet was held at the Panhandle College Cafeteria in 1940. The 4-H Federation was reorganized in 1941 and continues today.

In 1943, Texas County members earned more than $300,000. Mary Lee Sleeper, having sold the most Series E bonds, was selected to dedicate the Texas County B-24 Bomber and presented the Bomber to the Army Air Force at the state fair.

A camping trip earned by Joan Mires Chuesburg to Colorado began in the back of a semi cattle truck that had been washed with benches lining on both sides. "There were no backrests, but who cared!"

Currently, six 4-H clubs in Texas County have their roots in early 4-H clubs: Loyal Doers 4-H, chartered in 1961; Guymon 4-H; Yarbrough 4-H; Texhoma 4-H; Hardesty 4-H; Goodwell 4-H; and Cloverbud clubs in Goodwell, Hardesty, and Guymon.

Tillman County

Contributed by Micah Treadwell

Tillman County's 4-H history began in 1915, when the first county agent was appointed: W. A. Conner. Leona M. McKinley was appointed as the first home demonstration agent in 1917. A 1927 report showed sixteen 4-H clubs organized with 750 members. The same report had a comment that all of the clubs met during school hours with one-fourth of the day used for the 4-H meeting. The major projects were poultry, dairy, sewing, and canning. Perhaps the peak enrollment was in 1953; there were twenty-seven clubs and 1,411 members. Today, the population in our rural, southwest Oklahoma County is 9,287, which has dropped by 2,000 people in the last decade. However, the enrollment for 2009 was 374 members in ten traditional clubs and the shooting sports project club. Numbers have remained steady for several years. The most popular projects are livestock projects, especially swine and goats, photography, and shooting sports.

The 1950 4-H float. *Photo courtesy of the Tillman County Extension office.*

Tillman County has had its share of national and state winners. It would take much space to list all these winners, but we have had national project and judging events winners. Years ago, the county poultry judging team won at the national contest. Earning the most recent trips to national contests were the meat identification team (2007) and the shooting sports team (2006). Tillman County has had several district officers and one state 4-H president. Kenneth Holloway of rural Snyder served as state 4-H president from 1962 to 1963. Holloway

In April 2009, the Tillman County Land Judging Contest was held south of Frederick. *Photo courtesy of the Tillman County Extension office.*

In May 2007, the Outdoor Classroom at Hackberry Flat used the stream trailer. *Photo courtesy of the Tillman County Extension office.*

now resides southwest of Chattanooga. The earliest record of a national winner was in 1923, when Anna Mae Laurent of Tipton was selected to attend the International Livestock Show in Chicago. A National 4-H Congress may have been held then.

The county has seen change, but our 4-H mission is still the same: "To Make the Best Better."

Tulsa County

Contributed by Tracy Lane

Trucks and buses brought 103 4-H club members from Jenks to the Annual Club Rally in Tulsa, May 12, 1922. Several trucks had trouble staying on the washed-out county roads. The photograph was taken from the 1922 annual report of Home Demonstration Agent Katheryn E. Jackson. *Photo courtesy of the Tulsa County 4-H Program.*

Along the Arkansas River, on lands that were part of the Creek and Cherokee nations, Tulsa County began at statehood. The 4-H program developed a few years later. In 1922, 747 girls, enrolled in twenty-two clubs, participated in poultry, clothing, and horticulture projects. Currently, over 28,000 youth participate in clubs, school programs, special interest groups, and camps.

The 4-H clubs flourished during the forties and fifties, providing a sense of community and family in rural areas. In 1935, twenty-five active clubs held monthly meetings featuring business, demonstrations, and recreation. During World War I, members grew Victory Gardens extensively.

Gladys Logue, home demonstration agent, began recognizing 4-H achievement in 1948 with an afternoon tea. In 1954, Barbara Stunkard and Jerry Walker received the first Outstanding Boy and Girl Award. Changed to Hall of Fame

Paula Rowland, a Highland Park 4-H Club member, promotes National 4-H Club Week in 1973 at Southroads Mall in Tulsa. *Photo courtesy of Robert and Patsy Rowland.*

Tulsa County

Students get up close and personal with the sights and sounds of agriculture during a Petting Zoo at the YFR 4-H Club with 4-H member Johnathan Lane. *Photo courtesy of the YFR 4-H Club, Tulsa County.*

in 1984, it recognizes the two most outstanding members for progressive growth and leadership development.

Special federal funds appropriated in 1972 established urban 4-H programs. Derald Suffridge, the first urban agent, directed the work of five program assistants assigned to specific geographic areas reaching targeted low-income and minority youth. This program continues today in afterschool programs, Tulsa Housing Authority sites, and special interest clubs.

The groundwork of Charlene Nichols, home demonstration agent, in volunteer leadership enabled the development of phenomenal clubs including Dairy Capitol 4-H, leaders M. W. "Bud" and Willamae McAfee; Lynn Lane 4-H, leader Maxine "Granny" Baker; Valley View 4-H, leader Mrs. Russell Stunkard; and YFR 4-H, leaders Kyle and Carol Hunt.

The YFR club, chartered in 1982 as Young Farmers and Ranchers, was the vision of Clifford Vohon to keep teens positively directed by doing odd jobs around his farm. Five teenagers met in an old garage until Mr. Vohon provided the club with a four-acre plot of land. Here the teens worked together to build the 4-H clubhouse. Today, the club provides members the opportunity to raise and show animals and participate in other projects, too.

Tulsa County 4-H history is incomplete without the Tulsa County 4-H Foundation. Developing private financial support to 4-H from 1986 to 2006, charter members included David Mitchell, Leeland Alexander, and Joe Francis under the direction of Charlotte Richert, 4-H Extension educator.

Tulsa County 4-H. Because our future is too important to be left to chance.

In 2007, 4-H Youth Leadership Institute participants planned and executed a community service project for Life Senior Services that involved reviving their community garden and courtyard. *Photo courtesy of the Tulsa County 4-H Program.*

Wagoner County

Contributed by Jennifer Tresslar

Wagoner County youth who attended the 1999 State 4-H Roundup held on the OSU campus include Cody Conner, Emily Brunger, Lacy Conner, Paula Dean (county staff), Josh Keys, April Peck, Jason Stamps, Karen Blair, Tara Little, Brain Lamon, Alan Parnell (county staff), and Brandon Stamps.

Wagoner County can be found in the northeast section of Oklahoma, between Fort Gibson Lake and Tulsa County. The county was considered a front-runner in building its 4-H communities in Indian Territory. In the early days, Wagoner County was a booming agricultural community with cotton and livestock.

Wagoner serves as a gateway to Fort Gibson Lake for those coming from the West. A telegram was sent saying, "Wagoner's Switch is Ready" and in 1896, Wagoner became the first incorporated city in the Indian Territory (McMahan, 2010). In 1903, Wagoner had the first public school in the Indian Territory after Central College closed its doors in 1901 (McMahan, 2010). Wagoner was also known as the home of the famous racehorse Mr. Bar None.

Wagoner County's roots run deep in American Indian heritage. The county is located within the Creek Nation and the town of Porter was named after Chief Pleasant Porter, who led the Sequoyah Convention. The Sequoyah Convention focused on making Indian Territories separate from Oklahoma Territories (*Coweta American*, 2003). In a small community to the south of Porter sits Tullahassee. For many years, the Tullahassee Mission was considered the tallest building in the Indian Territory until it was destroyed by a fire in 1880 (*Coweta American*, 2003). Many Creek Freedom Students, the children of the black slaves of the Creeks who were freed after the Civil War, called the mission home (McMahan, 2010; *Coweta American*, 2003).

In October 1909, Wagoner County hired its first Cooperative Extension agent, Fred H. Ives, but after a few short months, Ives left the county office. Ives was replaced by C. N. Davis, who started on July 1, 1911, and held the position of county agent until March 31, 1913. Today, the Wagoner County Extension office houses four Extension educators and a nutrition education assistant. Wagoner County 4-H has seven clubs that serve 224 youth from urban and rural settings.

Wagoner County youth who attended the 2006 Northeast District Youth in Action Conference held at Fin and Father Resort at Tenkiller Lake are, L-R: Cody Richmond, Emily Sewell, Baylee Brown, Mandy Blair, Jena Parnell, and Ben Coffey.

This photograph provided by the Coweta American Archive shows the Tullahassee Mission with a group of Creek Freedom Students in 1890.

This photograph provided by the Coweta American Archive shows the importance cotton had in Wagoner County. In the county's early years, cotton was one of the major crops grown in Wagoner County.

Washington County

Contributed by Amy Berg

The very first 4-H club to be organized in Washington County was in September 1914, when five boys in the Cotton Valley Community formed the 4-H Corn and Kaffer Club. That was the beginning of 4-H clubs in Washington County, tutored by Mr. A. A. Powell, who was then the county agent. In 1917, Miss Ivy Burch joined the agent's staff, and in 1922, the first girls' 4-H Tomato Club was formed. In 1921, a county 4-H camp was held out at the old Boy Scout camp. For three days, 4-H boys and girls studied nature and were drilled on judging, demonstrations, and singing.

In 1943, every 4-H'er worked toward one goal. Through drives of all kinds, a B-24 Bomber was bought and named the Wash-co-homa, with the 4-H insignia painted on her side. At a special service at the Tulsa Airport, our 4-H Federation officers presented the Bomber to the United States Air Force.

Washington County 4-H members at Roundup. *Photo courtesy of the Washington County Extension office.*

The 4-H program continued to flourish from 1960 throughout the 1980s under the direction of Mr. George Seals. In 1991, Washington County 4-H helped with the Operation Desert Storm parade for the homecoming of soldiers of the Persian Gulf War in downtown Bartlesville. It was a big event and there were many soldiers who "walked" the parade and interacted with 4-H'ers.

In 1997, Operation Cow Chow was initiated, where 4-H volunteers and 4-H'ers rallied to round up eighty-four large round bales to send to farmers in drought-stricken southern Oklahoma. The 4-H'ers served coffee, juice, and doughnuts to workers and 4-H volunteers.

Washington County was lucky to have three outstanding young ladies as state 4-H ambassadors in the late 1990s and early 2000s. Janet Herren, Cathy Herren, and Kayla Swanson each served Washington County. Washington County 4-H is well known for consistently producing champion livestock judging teams under the direction of Randy Pirtle.

Washington County members explore all the interesting learning opportunities GPS has to offer. *Photo courtesy of the Washington County Extension office.*

Washita County

Contributed by Mary Peck

Lloyd Hawkins, Burns Flat, and his registered Chester White sow pig, which was bought for him in 1937 by his 4-H Club Department. Lloyd was to return to the Burns Flat 4-H Club two registered sow pigs at weaning time as compensation for the pig pictured. *Photo from the 1937 Annual Extension Report.*

The past one hundred years of 4-H has been a period of strong traditions and amazing changes. The program started with just a couple hundred members, reaching to more than 800 in the 1940s, and today enjoys the involvement of 400 youth. We boast many winners of state and national projects, state Hall of Fame recipients, and state 4-H officers. Our alumni are successful in all walks of life.

Many of the events held back then are still the 4-H events for today's 4-H'er. The first signs of 4-H activity surface in 1910 with 4-H Corn Club members exhibiting ten ears of corn in the county fair. In 1911, the Cordell Commercial Club, a precursor to the Cordell Chamber of Commerce, held a special contest for Corn and Pig Club members. Our first appropriate dress contest was held in 1930. First place won material and second place won seventy-five cents. Our first spring fat stock show was held in 1940. Winners of champion sheep and champion calf won trips to the American Royal Livestock Show. The winner of champion pig won a trip to National 4-H Congress in Chicago.

County 4-H camp is one of our big events. It's reported that the 1930 4-H camp was held at the Seger Indian Grounds at Colony with 164 delegates. The Cheyenne and Arapaho American Indians performed an Indian war dance for the evening program. Today, we hold 4-H camp at Camp Springlake south of Cordell.

The traditional projects such as clothing, food, gardening, livestock, and crops teach youth life skills just as they did one hundred years ago. Today's 4-H youth in Washita County begin the next one hundred years still strong in the projects of the past while adding the technology projects of the future such as computers, robots, GPS projects, and digital photography.

Washita County

Ida Katherin Okerberg (left) and Evelyn Davis (right) of Canute were winners of the Washita County Team Demonstration Junior Division in 1939. *Photo from the 1939 Annual Extension Report.*

The Washita County meat judging team judges meats at the Lawton Meat Supply Packing Company in 1950 in Lawton, Oklahoma. *Photo from the 1950 Annual Extension Report.*

Woods County

Contributed by Jill Whipple, Karen Armbruster, and Kourtney Coats

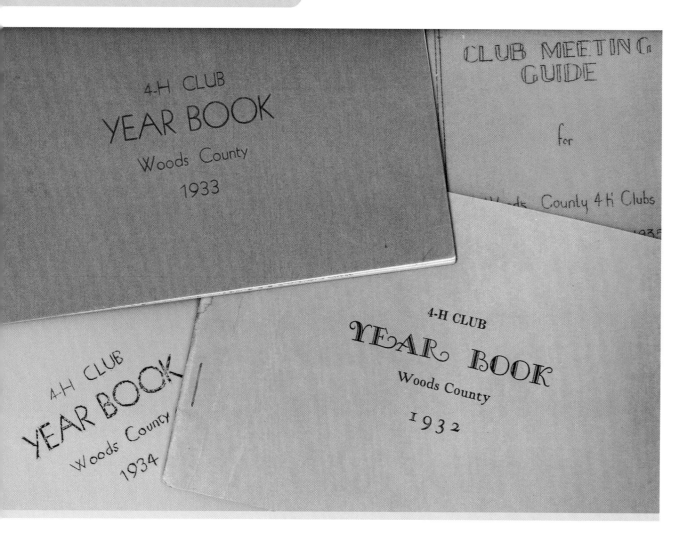

Record books have been a tradition in Woods County since the days of the Dust Bowl.

One of the earliest newspaper articles reported of the 1929 Woods County 4-H Girls' and Boys' Clubs getting started with their new year. The story reports that over 150 were in attendance at the meeting and a more lively and enthusiastic bunch was never seen.

The 1932 Woods County 4-H yearbook listed sixteen 4-H clubs. They included Capron, Riverview, Star, Friendship, Farry, Springdale, Lookout, Unity, Lone Star, Lone Star 73, Waynoka, Dacoma, Weber, Lake, and Franklin. All of these 4-H clubs were from communities throughout Woods County.

In 1935, Katherine McNally joined the Waynoka 4-H Club and became an active member. In a recent cell phone interview, Katherine stated that her parents made sure she and her sister Nettie attended the county 4-H meetings despite of the hardship of gas costs. Katherine was a state winner in the canning project area, and won a trip to Chicago by boarding a train in Ponca City, traveling to Kansas City, and then to Chicago. She said she was "raised to get in there and compete!" Katherine is now eighty-five years old and still competes with her beautiful cakes at the Woods County Fair.

Since 1952, sixteen Woods County 4-H members have received recognition in the Oklahoma Key Club. Since 1980, several members have been elected to serve on the Northwest District and as state 4-H officer. Those members include Amy Hofen, Scott Hofen, and Cassity Green. One of our 4-H members from the 1990s, Eric Stroud, served as Pistol Pete while attending Oklahoma State University!

Since 1926, there have been eighteen home demonstration agents, now known as Extension family and consumer science educators, and sixteen agriculture agents.

In 2009, there were three community clubs—Alva, Freedom, and Waynoka—with an enrollment of 130 members served by ten certified 4-H volunteers. There were also several project clubs in dog, photography, foods and nutrition, horse, and health.

Many 4-H'ers were rewarded trips for their top livestock projects. *Photo courtesy of the Woods County Extension office.*

Woodward County

Contributed by Melanie Matt

The earliest record of the Woodward County Extension Service's existence was March 25, 1914, when B. F. Markland served as the county agent. "A History of Woodward," published in *The Key Finder*, Volume III, No. 1 (January 1982) and Volume III, No. 2 (April 1982), written by Mrs. Mildred J. Hepner, stated that at this time the salary of the county agent was "supplemented with Chamber of Commerce funds."

It was in 1949 when Shirley Cooper Bedwell joined 4-H as a fourth-grade student. It was during her senior year of high school that Ms. Bedwell was named as Woodward County 4-H's first Hall of Fame recipient. During this time, Eugene Williams was serving as the county agent.

Ms. Bedwell said, through a telephone interview from her residence, which is still in Woodward County, that while she was a 4-H member, there were fourteen active 4-H clubs in Woodward County and each club had approximately twenty-five members. Her fondest 4-H memory was going to Stillwater to attend State 4-H Roundup. This 4-H trip proved valuable to her later as a student at Oklahoma State University.

Since that time, the number of clubs has become more concentrated. However, the number of 4-H members has seen a tremendous increase.

Oklahoma 4-H Hall of Fame

Year	Name	County
1950	Mae Audell Murray	Custer
	Bill Carmichael	Kay
1951	Joy Alexander	Washita
	R. J. Cooper	Woodward
1952	Nancy Brazelton	Kay
	Gordon Dowell	Payne
1953	Carolyn Crumm	Caddo
	Charles Chambers	Jefferson
1954	Virginia (Wyatt) Norman	Pottawatomie
	Zerle Carpenter	Greer
1955	Coleta McAlister	Kingfisher
	John Thornton	Washington
1956	Brenda (Adams) Tate	Payne
	Melvin Semrad	Garfield
1957	Bill Doenges	Washington
	Sharon Southard	Jefferson
1958	Bonnie (Fisher) Muegge	Payne
	Jimmy Quinn	Texas
1959	Doris (Wood) Smithee	Blaine
	Kenneth Schneeberger	Cotton
1960	Sue Dyson	Bryan
	Michael Lucas	Pontotoc
1961	Penny (Von Tungeln) Schwab	Canadian
	Kenneth Millican	Craig
1962	Mary Ann Hancock	Grady
	Chet Olson	Woods
1963	Frank Polach	Lincoln
	Mary Jean Grimes	Garfield
1964	Jaynee Murphy	Oklahoma
	Larry Quinn	Texas
1965	Ann Williams	Logan
	Earl Cinnamon	Garfield
1966	Nancy (Stiles) Henry	Payne
	Pete Teel	Texas
1967	Jerry Harke	Texas
	Glenda Estill	Garfield
1968	Garvin Quinn	Texas
	Sue (Gunkel) Paden	Jackson
1969	Sharon Pospisil	Garfield
	Mike Synar	Muskogee
1970	John Harp	Delaware
	Dixie Shaw	Alfalfa
1971	Clayton Taylor	Muskogee
	Latriece (Baker) Connolly	Beckham
1972	John Roush	Alfalfa
	Penny (Kruska) Hook	Greer
1973	Rodd Moesel	Oklahoma
	Pam Miller	Garfield
1974	Tammie Morrison	Payne
	Joe Francis	Cleveland
1975	Gwen Shaw	Alfalfa
	Mike Nelson	Beckham
1976	Melinda Shockey	Grady
	James Pfeiffer	Logan
1977	Linda Hart	Lincoln
	Mike Barrington	Okfuskee
1978	Mark Detten	Kay
	Sheila Alexander	Canadian
1979	Becky (Krittenbrink) Schnaithman	Grant
	Rene Crispin	Dewey
1980	Kent McVey	Payne
	Mary Myles (Detten) Radka	Kay
1981	Renetta (Reeves) Bullard	Texas
	Paul Mackey	Beckham
1982	Marla (Johnson) Barnes	Ellis
	Kirk Smithee	Oklahoma
1983	Dana Bock	Greer
	Toby Don Wise	LeFlore
1984	Matt Calavan	Muskogee
	Danny Mackey	Beckham
1985	Nancy Jo (Smith) Howard	Garvin
	Lavonna (Wade) Hopkins	Logan
1986	Connie Jo Bierig	Blaine
	Jonathan Kolarik	Pottawatomie
1987	Vernon McKown	Cleveland
	Janeen (Peters) McGuire	Tulsa
1988	Natalie (James) Church	Grant
	Jerry D. Kiefer	Caddo
1989	Angela (Howard) Schuster	Cleveland
	Dana (Spurgeon) Chaffin	Tulsa
1990	Keith Tracy	Roger Mills
	Gary Piercey	Washita
1991	John Tracy	Roger Mills
	Tara Sherrer	Beaver

Year	Name	County
1992	Alana Malone	Custer
	Krissinda (Reece) Williams	LeFlore
1993	Chuck Lester	Noble
	Chereece (Currington) Jackson	Jefferson
1994	Josh London	Tulsa
	Janice Riley	Comanche
1995	Lindsay Sherrer	Beaver
	Shannon Ferrell	Roger Mills
1996	Jill (Anderson) Rucker	Payne
	Charla Dobson	Custer
	Jerry Melichar	Tulsa
	Sarah (Barton) Lane	Creek
1997	Missy Conner	Garfield
	Kent Gardner	Woodward
1998	Ryan McMullen	Washita
	Matt Barton	Creek
1999	Marcy (Grundmann) Luter	Pottawatomie
	Derrick Ott	Cleveland
2000	Jim Rexwinkle	Ottawa
	Steven Anderson	Payne
2001	Leigh Ann Tracy	Garvin
	Rachel (Keeling) Moehle	Lincoln
2002	Caleb Winsett	Pottawatomie
	Will McConnell	Lincoln
2003	Josh Grundmann	Pottawatomie
	Dusty Conner	Garfield
2004	Kirby Teachey	Pontotoc
	Magan Smith	Pontotoc
2005	Cathleen Taylor	Pontotoc
	Nathan Thompson	Lincoln
2006	Jered Davidson	Caddo
	Tiffany Grant	Tulsa
2007	Jerret Sanders	Caddo
	Carrie Highfill	Garfield
2008	Meg McConnell	Lincoln
	Matthew Taylor	Pontotoc
2009	Julie Bragg	Cleveland
	Kylie Stowers	Cleveland

Oklahoma 4-H State Presidents

Year	Name	County
1929–1930	Arthur Sweet	Greer
1930–1931	Odes Harwood	
1931–1932	Burl Winchester	Garfield
1932–1933	Carl Neuman	Greer
1933–1934	Henry Osburn	Pottawatomie
1934–1935	Edgar McVicker	Roger Mills
1935–1936	Allen Goodbary	Lincoln
1936–1937	Arnold Neuman	Greer
1937–1938	Bob Morford	Alfalfa
1938–1939	Harry Synar	Muskogee
1939–1940	Wilburn Wiley	Okfuskee
1940–1941	Aaron Gritzmaker	Garfield
1941–1942	Dayton Rose	Okfuskee
1942–1943	Edmond Synar	Muskogee
1943–1944	Robert Lee Nash	Jefferson
1944–1945	Henry Wolf	Hughes
1945–1946	Valentino Synar	Muskogee
1946–1947	Gerald Honick	Kay
1947–1948	Ernest Hellwege	Kingfisher
1948–1949	Ted Davis	Kiowa
1949–1950	Don Bliss	Kay
1950–1951	Charles Smith	Pittsburg
1951–1952	Bill Charmichael	Kay
1952–1953	Charles Chambers	Jefferson
1953–1954	Dickie Yates	Beaver
1954–1955	Tom Tate	Osage
1955–1956	Melvin Semrad	Garfield
1956–1957	Don Davidson	Tulsa
1957–1958	James Quinn	Texas
1958–1959	Keith Smith	Oklahoma
1959–1960	Mike Lucas	Pontotoc
1960–1961	David Semrad	Garfield
1961–1962	Bruce Robinett	Garfield
1962–1963	Kenneth Hollaway	Tillman
1963–1964	John Hancock	Grady
1964–1965	Jim Loepp	Beaver
1965–1966	Chuck Daughtery	Tulsa
1966–1967	Richard Bailey	Pontotoc
1967–1968	Larry Brooks	Oklahoma
1968–1969	Dee Griffin	Seminole

Year	Name	County
1969–1970	John Harp	Delaware
1970–1971	Clayton Taylor	Muskogee
1971–1972	Gary Stone	Seminole
1972–1973	Morris Strom	Noble
1973–1974	Ken McQueen	Murray
1974–1975	Joe Francis	Cleveland
1975–1976	Gary Marshall	McCurtain
1976–1977	Steve McCarter	Muskogee
1977–1978	Mark Detten	Kay
1978–1979	Rene Crispin	Dewey
1979–1980	Terry Henderson	Muskogee
1980–1981	Paul Mackey	Beckham
1981–1982	Marla Johnson	Ellis
1982–1983	Toby Don Wise	LeFlore
1983–1984	Matt Calavan	Muskogee
1984–1985	Rodger Kerr	Jackson
1985–1986	Kent Major	Blaine
1986–1987	Vernon McKown	Cleveland
1987–1988	Jerry Kiefer	Caddo
1988–1989	Natalie James	Grant
1989–1990	Keith Tracy	Roger Mills
1990–1991	Amy Miller	Muskogee
1991–1992	Chuck Lester	Noble
1992–1993	Donna Quinn	Comanche
1993–1994	John Cothren	Garvin
1994–1995	Shannon Ferrell	Roger Mills
1995–1996	Paula Moore	Caddo
1996–1997	Jonathan Smith	Stephens
1997–1998	Missy Conner	Garfield
1998–1999	Ryan McMullen	Washita
1999–2000	Jim Rexwinkle	Ottawa
2000–2001	Rodd Holland	Tulsa
2001–2002	Caleb Winsett	Pottawatomie
2002–2003	Dusty Conner	Garfield
2003–2004	Jennifer Long	Muskogee
2004–2005	Kirby Teachey	Pontotoc
2005–2006	Jered Davidson	Caddo
2006–2007	Laramy Wilson	Hughes
2007–2008	Jerret Sanders	Caddo
2008–2009	Matt Taylor	Pontotoc
2009–2010	Sam Durbin	Canadian

Oklahomans in the National 4-H Hall of Fame

Year	Name
2002	Mary Sue Sanders
2003	Eugene "Pete" Williams
2004	Ernest Holloway
2006	Ray Parker
2008	Ira J. Hollar
2009	Barbara Hatfield
2009	Joe Hughes

Oklahoma 4-H State Leaders

Year	Name
1901–1911	W. D. Bentley, father of Extension in Oklahoma
1909	Corn and Tomato Clubs formed
1911–1912	Fred Ives
1912–1913	N. E. Winters
1912	Mrs. A. E. Walker appointed first Home Demonstration Agent
1913	Work moved from Yukon to Oklahoma City Federal Building
1913–1914	T. M. Jeffords
1914	Extension work moved to Oklahoma A&M College
1914–1915	James A. Wilson
1915–1922	John E. Swain
1922–1924	E. B. Shotwell
1924–1939	B. A. Pratt
1940–1949	Paul G. Adams
1950–1965	Ira Hollar, State 4-H Leader
1966–1980	Eugene "Pete" Williams, Assistant Director and 4-H Leader
1980–1988	Wallace O. Smith, Assistant Director and 4-H Leader
1989–1994	James A. "Jim" Rutledge, Assistant Director and 4-H Leader
1994–1996	Charles B. Cox, Interim Program Leader
1996–1997	Fred Rayfield, Program Leader
1997–2007	Lynda Harriman, Assistant Director FCS and 4-H
1998–2007	Charles B. Cox, Program Leader
2007–present	Charles B. Cox, Assistant Director and 4-H Leader

Bibliography

County Extension Office 4-H Records.

Coweta American, 2003.

Daily Ardmoreite, January 20, 1922.

Goodwin, Sam, et al. Oral history interviews, 2009.

Grant County Extension Office. The 4-H Yearbook, July 1948.

Hepner, Mildred J. "A History of Woodward." *The Key Finder*, Vol. III, No. 1 (January 1982) and Vol. III, No. 2 (April 1982).

"Loretta Bauer Tops Winners In 4-H Work." *Guymon Daily Herald*, November 24, 1957.

Marshall County Genealogy Historical Society Records.

"Marshall Henderson," *The Republican*, October 19, 1928, Section 1, p. 1.

McConnell, F. B. Letter to Mary Gabrish, February 21, 1950.

McMahan, Liz. "Welcome to Wagoner—Where a Century Has Taken Us… From Queen City of Prairies… To Gateway to Fort Gibson Lake." Wagoner History Presentation, 2010.

Meredith. "Advice to Farm Boys and Girls," *Jet Visitor*, July 8, 1920, Section 1, p. 1.

Oklahoma County Agriculture Agents. Annual Narrative Report, 1926, 1927, 1929, 1930, 1937, 1939, 1940, 1941, and 1949.

Oklahoma State University Special Collections and University Archives.

Vinita Daily Journal, May 5, 1989.

About the Authors

Jessica Stewart began her career with Oklahoma 4-H in 2008 as the coordinator of special programs and promotions. Although she had no previous experience with 4-H Youth Development Programs, Jessica has enjoyed celebrating the Oklahoma 4-H Centennial with youth, volunteer leaders, alumni, and donors. Jessica says she is excited about what 4-H will do in the next one hundred years, and feels privileged to have worked with so many dedicated people involved in the program. Jessica has a Bachelor of Science from Oklahoma State University in agricultural communications. In her spare time, Jessica enjoys spending time with her husband, David, and their horses and dogs.

Caitlin Scheihing is a native Oklahoman, born and raised in Greenfield. Caitlin joined 4-H when she was nine and was involved for four years. Her projects focused on livestock production and public speaking. Caitlin began her position at the state 4-H office in the spring of 2009 as a marketing and communications intern. She then took over the position of centennial history book editor in August. Caitlin says she has enjoyed her time at the state 4-H office as it has given her the opportunity to gain a new perspective on Oklahoma 4-H programs. Caitlin has a Bachelor of Science from Oklahoma State University in agricultural communications with a minor in agricultural leadership. She plans to become a 4-H club leader in her community so that she can help others learn about 4-H and the leadership skills it has to offer.